Mathematics for Health Professionals

P9-CJJ-916

Mathematics for Health Professionals

Second Edition

B. Louise Whisler
San Bernardino Valley College

 Wadsworth Health Sciences Division
Monterey California

Wadsworth Health Sciences Division
A Division of Wadsworth, Inc.

Printed in the United States of America

10 9 8 7 6 5 4 3 2 1

Library of Congress Cataloging in Publication Data

Whisler, B. Louise, [date]–
  Mathematics for health professionals.

  Bibliography: p.
  Includes index.
  1. Nursing—Mathematics.   2. Nursing—Mathematics—
Problems, exercises, etc.   3. Medicine—Mathematics.
4. Medicine—Mathematics—Problems, exercises, etc.
I. Title.
RT68.W45   1985   510'.2461   84-21911
ISBN 0-534-04350-X

Sponsoring Editor: Aline Faben
Editorial Assistant: Mary Beth McDavid
Production Editor: Constance D. Brown
Permissions Editor: Mary Kay Hancharick
Interior and Cover Design: Vargas/Williams/Design
Art Coordinator: Judith Macdonald
Interior Illustration: Carl Brown
Typesetting: Omegatype Typography, Inc., Champaign, Illinois
Printing and Binding: Malloy Lithographing, Inc., Ann Arbor, Michigan

Preface

Mathematics for Health Professionals is the product of many years' development and classroom testing at San Bernardino Valley College. The nursing department instructors helped with and guided the book's development; the mathematics department instructors assisted with the testing.

The book is constructed to perform three major services for those in the health science professions. The first is to review the arithmetic and algebra that students need for calculations of health-related problems. It is especially concerned with developing comprehension, accuracy, good checking methods, and retention of the necessary mathematical skills.

The second service is to provide a sufficiently comprehensive presentation of the various measurement systems used in health-related occupations. These systems of measurement include the SI (metric system), the apothecaries' system, the ordinary U.S. system, and the household system. In addition, there is a chapter devoted to the study of drugs measured in units.

The third service is to provide a detailed discussion of the specific types of problems performed by those in health-related occupations. These include medication dosage problems of all degrees of complexity, solution-labeling comprehension problems, solution preparation problems, IV administration problems, pediatric dosage problems, and burn patient problems. In addition, the use of the various measurement systems in medication problems is discussed at length.

Mathematics for Health Professionals may be used for regular classroom instruction or for self-study. In either case, for success in use of this book a good working knowledge of arithmetic operations using whole numbers is required at course entry. The instructor is provided a placement test with suggested score range for course admission based on research findings. Also provided are sample examinations, including the final, and a complete solution manual.

A brief look at the book's structure may be helpful. The first three chapters review arithmetic operations, and each includes a pretest to help the instructor determine what, if any, material needs to be studied and where emphasis is needed. These three chapters contain much material to give the students confidence in computations and to make calculations easier. In this edition there is additional work with percentages, and changes have been made that reflect findings in classroom research aimed at increasing understanding.

Chapter 4 is a key chapter in the text. This chapter introduces the algebra required for medical calculations, and contains the development of the basic solution pattern for the vast majority of medication problems. The proportion is the mathematical tool for solution of all but a few of the medication problems that health science students will be responsible for solving. Introduced here, the proportion is used with great frequency in the remainder of the book. It is hoped that the student will become comfortable, skilled, and accurate in solution of proportions. Chapter 4 also includes the first formal dosage calculation problems. These problems continue in ever-increasing complexity in the remainder of the book.

Chapters 5, 6, and 7 present the apothecaries', metric, U.S. Customary, and household units of measure, as well as conversion possibilities from one to the other. The metric system symbols and spellings conform to the International System of Units standard, which is the international standard used by all nations and which is recommended by the U.S. Bureau of Standards for all college level work. Chapter 8 includes the interpretation of solution labels and two optional sections on preparation of solutions. Chapters 9, 10, and 11 contain the specialized topics of drugs measured in units, drugs administered intravenously, and pediatric dosages. Chapter 12, newly written for this edition, explains estimation of percentage of body surface burned in the burn patient. If the instructor wishes, Chapter 12 may be done following Chapter 3.

You will find that Chapters 4 through 12 contain a wealth of medication problems. With the passage of time, the availability of a given drug or the optimal dosage of a given drug may change when results from research and/or clinical observations are evaluated. So, periodically, a text such as this must be updated to reflect these changes. Utilizing the 1984 *American Hospital Formulary Service,* this update has been completed and each drug and dosage in this edition is current at the time of this writing. Body measurements found in the book are from Gray's *Anatomy of the Human Body.*

Use of certain equipment has been found to increase understanding of the material in Chapters 4 through 11. Therefore, the author recommends that each student be given a plastic 1-ounce measuring cup (see Figure 7.1) when Chapter 7 is studied. Also useful are classroom size sets of metric rulers, metric tapes, cubic centimetres, tuberculin syringes, 2.5-ml syringes, and insulin syringes. Additional recommended equipment includes a 1-l flask, a 100-ml graduated cylinder, some gram mass pieces, some meter sticks, and a metric bathroom scale.

Finally, I wish to thank the following people without whose help the book could not have been written. From San Bernardino Valley College: Betty Jacobson, the former health division chairperson, who provided the structure of the material and the original encouragement; Dorothy Scantlin, health division chairperson; Eileen Battle, Betty Vargas and other nursing instructors, who provided continuing advice and information; Robert Wakefield, mathematics instructor, who suffered through and helped with the endless revisions of material. Jan Earnhart, Loma Linda University School of Nursing, who provided guidance on the burn patient chapter content and resource materials. Jack Porter, Cuyahoga Community College, Dane Lapkin, Middlesex Community College, and Ellen B. Gloyd, Montgomery College, who thoughtfully reviewed various manuscript drafts; Martha Kadow, whose professional typing and knowledge of manuscript requirements made my job easier; and my husband, Donald Whisler, whose patience and understanding made the whole project possible.

B. Louise Whisler

From the Author to the Student

At course entry, *Mathematics for Health Professionals* requires a good working knowledge of arithmetic operations using whole numbers. Meeting this condition, the student may study the book in a regular classroom setting or use the book for self-study. For those students wishing to pursue self-study, some comments should be made about the first three chapters since all students will have prior exposure to most of the material. It is advised that the student take the pretests and work the chapter reviews to determine which parts of these three chapters should be given extra attention. All sections of these chapters should be reviewed by all students to refresh memory and to pick up operational techniques to improve accuracy. Close attention to these introductory chapters will make work easier in later chapters.

There are many text features to assist you in your study: chapter reviews are included to help you summarize material completed or review for examinations; answers to odd-numbered problems are found at the end of the book; a complete solution manual is available; there are lots of examples in each chapter; and the appendixes contain valuable reference materials. To obtain greatest advantage, become familiar with the appendixes contents early in the course.

Students who have completed the study of *Mathematics for Health Professionals* satisfactorily have shown good retention and a high degree of mathematical accuracy in subsequent nursing and other health science classes. They also have scored high on mathematics placement tests.

Finally, you may want to read the preface to get an idea what your author has in mind for you. I am frequently asked "What degree of accuracy do you expect?" I must admit I want you to be as nearly perfect as possible. Good luck and do enjoy!

Contents

Mathematics for Health Professionals

Common Fractions

Since common fractions are used extensively for solving medication problems, this first chapter includes a detailed review of common fraction solution skills. In some cases you may be asked to think in a new or different manner in order to attain efficiency and accuracy.

Topics include:

Ways of writing, reading, and interpreting the common fraction.

Number theory to make your work easier

Operations of addition, subtraction, multiplication, and division involving common fractions

Pretest   [1.1]

1. In $\frac{2}{3}$ the position held by the 2 is called the _____ or the _____ .

2. In $\frac{2}{3}$ the position held by the 3 is called the _____ or the _____ .

[1.2]

3. Using six circles, illustrate $\frac{2}{3}$.

4. What is the prime factorization of 18?

[1.3]

5. What is the prime factorization of 247?

6. State the Identity Property of Multiplication.

7. $\dfrac{7}{8} = \dfrac{?}{184}$

8. Reduce $\frac{9}{45}$ to lowest terms.

[1.4]

9. Reduce $\frac{221}{391}$ to lowest terms.

10. The least common multiple of 9, 15, and 20 is _____ .

[1.5]

11. $\dfrac{1}{9} + \dfrac{2}{3} + \dfrac{3}{10} =$

12. $\dfrac{2}{9} + \dfrac{3}{20} - \dfrac{4}{15} =$

[1.6]

13. Change $\frac{28}{6}$ to a mixed number.

14. Change $3\frac{4}{5}$ to an improper fraction.

15. $2 + 3\dfrac{7}{8} + 4\dfrac{1}{3} =$

16. $3\dfrac{1}{7} - 2\dfrac{3}{4} =$

[1.7]

17. $\dfrac{5}{42} \cdot \dfrac{14}{33} \cdot \dfrac{22}{35} =$

18. $1\dfrac{3}{4} \cdot 2\dfrac{7}{8} =$

19. What is $\frac{1}{3}$ of $2\frac{2}{5}$?

20. $1\dfrac{3}{4} \div 2\dfrac{7}{8} =$

21. $\dfrac{7}{10} \div 100 =$

22. $\dfrac{7}{10} \div \dfrac{1}{100} =$

23. $\frac{1}{100}$ is how many times $\frac{1}{200}$?

24. $\frac{1}{6}$ is what part of $3\frac{2}{3}$?

[1.8]

25. $\dfrac{\frac{1}{150}}{\frac{1}{100}} =$ 26. $\dfrac{2\frac{1}{3}}{\frac{1}{3}} =$

[1.9]

27.   Insert $=, >$, or $<$ between the fractions to make them true statements.

  a. $\dfrac{1}{100} \quad \dfrac{1}{200}$ b. $\dfrac{8}{13} \quad \dfrac{11}{17}$

What They Mean

Section 1.1

The common fraction is a comparison of two numbers, a and b, commonly written: $\frac{a}{b}$ in which the value of a and b can be any number, but b cannot be zero since division by zero has no meaning. Other forms equivalent to $\frac{a}{b}$ are

Ways of Writing Common Fractions

$$a \div b \qquad a : b \qquad b\overline{)a} \text{ (used for long division)}$$

The common fraction $\frac{a}{b}$ may be read in many different ways, including

Ways of Reading Common Fractions

$$\frac{a}{b} \qquad\qquad \frac{3}{4}$$

a divided by b	3 divided by 4
a is to b	3 is to 4
the part a divided by the whole b	the part 3 divided by the whole 4
b divided into a	4 divided into 3

There are different names for the positions occupied by a and b. Some of these are

Positional Names

$$\frac{\text{numerator}}{\text{denominator}} \qquad \frac{\text{part}}{\text{whole}} \qquad \frac{\text{dividend}}{\text{divisor}}$$

You should remember these names.

The Part–Whole Relationship   Using the common fraction $\frac{3}{4}$, let's look at the part–whole relationship. The denominator position in the fraction $\frac{3}{4}$ tells us to separate the whole amount of something into 4 equal parts (Figure 1.1).

Figure 1.1

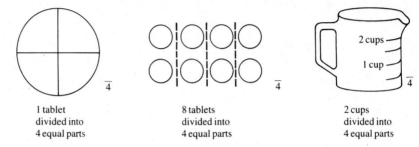

1 tablet
divided into
4 equal parts

8 tablets
divided into
4 equal parts

2 cups
divided into
4 equal parts

The numerator in the fraction $\frac{3}{4}$ tells us to consider 3 of the 4 equal parts (Figure 1.2).

Figure 1.2

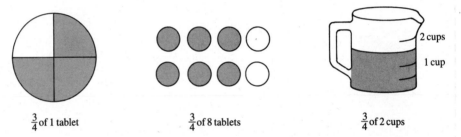

$\frac{3}{4}$ of 1 tablet $\frac{3}{4}$ of 8 tablets $\frac{3}{4}$ of 2 cups

An interesting situation occurs when the numerator of a fraction is larger than its denominator. To understand, use the concept we just discussed and the fraction $\frac{5}{4}$. In Figure 1.3 we see that another name for $\frac{5}{4}$ (five-fourths) is $1\frac{1}{4}$ (one and one-fourth). A fraction that has a numerator larger than its denominator is called an *improper fraction*. We will discuss improper fractions at length in Section 1.6.

Figure 1.3

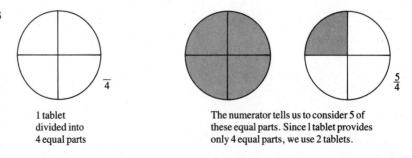

1 tablet
divided into
4 equal parts

The numerator tells us to consider 5 of
these equal parts. Since 1 tablet provides
only 4 equal parts, we use 2 tablets.

Again, let's use $\frac{3}{4}$ to illustrate. In this relationship $\frac{3}{4}$ is read *3 is to 4.* The fraction simply compares the number 3 with the number 4, and may be written $3:4$ as well as $\frac{3}{4}$. Both forms may be read *3 is to 4.* We will discuss the ratio form in Chapter 4.

The Ratio Relationship

In the third relationship we consider $\frac{3}{4}$ to be $3 \div 4$ or $4\overline{)3}$. A quantity of 3 is divided into 4 equal parts. For example, 3 dozen eggs divided into 4 equal parts would look something like Figure 1.4.

The Division Relationship

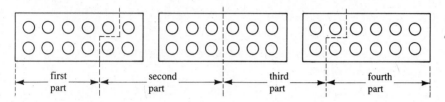

Figure 1.4

In summation $\frac{a}{b}$ is equivalent to $a \div b$, $a:b$, and $b\overline{)a}$. All four forms indicate the operation of division, and *a* is the *dividend,* *b* is the *divisor,* and the *quotient,* in each case, is the answer to the division problem.

$$\text{divisor} \overline{)\text{dividend}}^{\text{quotient}}$$

To solve medication problems, you will have to know the various ways of reading common fractions and their various meanings.

1. Using six tablets, illustrate each of the following common fractions:

    a. $\dfrac{1}{3}$ b. $\dfrac{2}{3}$ c. $\dfrac{1}{2}$ d. $\dfrac{1}{6}$

Exercises 1.1

2. Use Exercise 1 to help you with this problem. The doctor asks you to give your patient six tablets each day; one-half after breakfast, one-third after lunch, and the rest after dinner.

    a. How many tablets will you administer after breakfast?

    b. How many tablets will you administer after lunch?

    c. How many tablets will you administer after dinner?

3. You have 10 ounces liquid. Illustrate the following:

    a. $\dfrac{2}{5}$ b. $\dfrac{7}{10}$ c. $\dfrac{4}{5}$ d. $\dfrac{1}{2}$

4. You have a 10-ounce container of liquid. You've been asked to give your patient three-fifths of the liquid.

   a. How many ounces will you administer?

   b. How many ounces will you have left?

5. Name the common fractions illustrated in each shaded area.

   a. 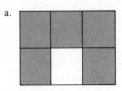   b. 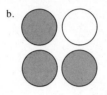   c.

6. You have three tablets that must be divided into six equal doses. Illustrate.

7. You have two tablets that must be divided into three equal doses. Illustrate.

8. You began with 9 millilitres penicillin and have administered 2 millilitres. Express as a common fraction.

   a. the part used.   b. the part remaining.

Section 1.2   **Number Theory**

Before we begin to work with fractions, we should review some basics of number theory and mathematical operation.

A number is divisible by another number if the operation of division results in no remainder. For example, 246 is divisible by 2 because 246 divided by 2 equals 123 with no remainder; 246 is divisible by 3 because 246 divided by 3 equals 82 with no remainder. But 246 is not divisible by 4 because 246 divided by 4 equals 61 with a remainder of 2.

Tests for Divisibility   Of course, we can perform the division to see if one number is divisible by another number, but there are shortcuts. The following are three frequently used tests for divisibility.

1. *A number is divisible by 2 if the final digit of the number is 0, 2, 4, 6, or 8.*

    135 796 is divisible by 2 because the final digit on the right is 6. On the other hand, 24 683 is not divisible by 2 because its final digit is 3. (Check these by division.)

2. *A number is divisible by 3 if the sum of the digits in the number is divisible by 3.*

    14 622 is divisible by 3 because $1 + 4 + 6 + 2 + 2 = 15$ and $15 \div 3 = 5$ with no remainder. The sum of the digits (15) is divisible by 3; therefore, the number (14 622) is divisible by 3. (Check.)

    24 611 is not divisible by 3 because $2 + 4 + 6 + 1 + 1 = 14$ and 14 is not divisible by 3. (Check.)

3. *A number is divisible by 5 if the final digit is 0 or 5.*

    145 is divisible by 5 because the final digit is 5.

    506 is not divisible by 5 because the final digit is 6.

There are many other tests for divisibility, but you will find these three the most frequently used.

Prime and Composite Numbers

A prime number is any number that is divisible only by itself and one. Two is a prime number because 2 is divisible by itself and 1, but by no other number; 7 is a prime number because 7 is divisible only by itself and 1. You should memorize these first ten prime numbers:

$$2 \quad 3 \quad 5 \quad 7 \quad 11 \quad 13 \quad 17 \quad 19 \quad 23 \quad 29$$

A composite number is any number that is divisible by some number other than itself and one. Four is a composite number because 4 is divisible by 4, by 1, *and* by 2. Six is a composite number because 6 is divisible by 6, by 1, *and* by 3 and 2. The first ten composite numbers are

$$4 \quad 6 \quad 8 \quad 9 \quad 10 \quad 12 \quad 14 \quad 15 \quad 16 \quad 18$$

All counting numbers with the exception of 1, which is in a class by itself, are either prime numbers or composite numbers.

Prime Factorization *A factor of a number is a number that is an exact divisor of the given number.* Four is a factor of 12 because 12 is divisible by 4. In fact, 1, 2, 3, 4, 6, and 12 are all factors of 12.

> 1 and 12 are factors of 12 because $1 \cdot 12 = 12$.
>
> 2 and 6 are factors of 12 because $2 \cdot 6 = 12$.
>
> 3 and 4 are factors of 12 because $3 \cdot 4 = 12$.

In each of these examples at least one factor is a composite number. Let's look at one more set of factors:

> $2 \cdot 2 \cdot 3$ is called the *prime factorization of* 12

In a *prime factorization* each of the factors (and they may be repeated) is a prime number. The *prime factored form of 18 is* $2 \cdot 3 \cdot 3$ *because each factor is a prime number and their product is 18.*

Every composite number has one and only one prime factorization. In other words, *the prime factored form of any composite number is unique.*

Since we use prime factors to work with fractions, we need an efficient method for obtaining the prime factorization of any given number. *To find the prime factorization of a number, test the divisibility of the number by the prime numbers in the order in which the primes occur.* Make sure that all factors of any one prime are divided out before you test the quotient's divisibility by the next prime. Proceed in this manner until the final quotient is a prime number. *The final quotient and the prime number divisors are the prime factors of the given number.*

EXAMPLE 1 Find the prime factorization of 30.

1. Is 30 divisible by 2? Yes.

   $$30 \div 2 = 15$$

   2 is a prime factor of 30.
2. Is 15 divisible by 2? No. Proceed to the next prime number.
3. Is 15 divisible by 3? Yes.

   $$15 \div 3 = 5$$

   3 is a prime factor of 30.

4. The final quotient is 5, which is a prime number so $2 \cdot 3 \cdot 5$ is the prime factorization of 30.

EXAMPLE 1
continued

The work pattern for Example 1 would look like this:

$$
\begin{array}{r}
5 \\
3\overline{)15} \\
2\overline{)30}
\end{array}
$$

You can see that each divisor and the final quotient make up the prime factorization. The method works easily, even with larger numbers.

Find the prime factorization of 693.

EXAMPLE 2

1. Is 693 divisible by 2? No. Proceed to the next prime number.
2. Is 693 divisible by 3? Yes.

    $$693 \div 3 = 231$$

    3 is a prime factor of 693.
3. Is 231 divisible by 3? Yes, so divide by 3 again.

    $$231 \div 3 = 77$$

    A second 3 is a prime factor of 693.
4. Is 77 divisible by 3? No. Proceed to the next prime number.
5. Is 77 divisible by 5? No. Proceed to the next prime number.
6. Is 77 divisible by 7? Yes.

    $$77 \div 7 = 11$$

    7 is a prime factor of 693.
7. The final quotient is 11, which is a prime number so $3 \cdot 3 \cdot 7 \cdot 11$ is the prime factorization of 693.

The work pattern for Example 2 would look like this:

$$
\begin{array}{r}
11 \\
7\overline{)\ 77} \\
3\overline{)231} \\
3\overline{)693}
\end{array}
$$

Again, you can see that each divisor and the final quotient make up the prime factorization.

Exercises 1.2   In Exercises 1–8, state whether each number is divisible by 2, by 3, or by 5.

1.	42	2.	57
3.	123	4.	3298
5.	17	6.	180
7.	165	8.	48 296

9.   List the first 12 prime numbers.

Find the prime factorization of each of the following numbers. Be sure to show your work.

10.	35	11.	15
12.	16	13.	18
14.	100	15.	83
16.	105	17.	24
18.	483	19.	13 860
20.	421	21.	62
22.	1182	23.	37
24.	201	25.	667
26.	91	27.	9
28.	12	29.	87

Section 1.3   Equivalent Fractions

To perform arithmetic operations with common fractions, we frequently must replace a common fraction with an equivalent fraction. The replacement may have larger numbers (building) or smaller numbers (reducing). In either case we obtain the desired result by a special application of the Identity Property of Multiplication.

The *Identity Property of Multiplication* states that the product of **Identity Property**
1 and any given number (whole number or fraction) is equal to the **of Multiplication**
given number. That is,

$$a \cdot 1 = 1 \cdot a = a$$

When applying the Identity Property of Multiplication, remember that 1 has many different names:

$$\frac{2}{2}, \frac{4}{4}, \frac{15}{15}, \frac{176}{176}$$

Each is a name for 1. So,

$$\frac{2}{3} \cdot 1 = 1 \cdot \frac{2}{3} = \frac{2}{3}$$

and

$$\frac{2}{3} \text{ (given fraction)} \cdot \frac{4}{4} \text{ (multiplication by 1)}$$

$$= \frac{8}{12} \left(\text{equivalent to } \frac{2}{3} \right)$$

Above, when we multiplied $\frac{2}{3}$ by $\frac{4}{4}$, our name for 1, we built the **Building**
equivalent fraction $\frac{8}{12}$ using the Identity Property of Multiplication. **Equivalent**
To *build an equivalent fraction*, convert the original fraction to an **Fractions**
equal-in-value fraction that has a larger denominator by multiplying by 1.

Given the fraction $\frac{2}{5}$, find an equivalent fraction with a denomi- **EXAMPLE 1**
nator of 20, or

$$\frac{2}{5} = \frac{?}{20}$$

To begin, think "5 times what is 20?" Since the answer is 4, the
name for 1 needed is $\frac{4}{4}$.

$$\frac{2}{5} = \frac{2}{5} \cdot \frac{4}{4} = \frac{8}{20}$$

$\frac{8}{20}$ is the fraction equivalent to $\frac{2}{5}$ with a denominator of 20.

EXAMPLE 2

$$\frac{5}{18} = \frac{?}{180}$$

$$= \frac{5}{18} \cdot \frac{10}{10}$$

$$= \frac{50}{180}$$

EXAMPLE 3

$$\frac{1}{12} = \frac{?}{180}$$

$$= \frac{1}{12} \cdot \frac{15}{15}$$

$$= \frac{15}{180}$$

EXAMPLE 4

$$\frac{2}{15} = \frac{?}{180}$$

$$= \frac{2}{15} \cdot \frac{12}{12}$$

$$= \frac{24}{180}$$

Reducing Common Fractions We also use the Identity Property of Multiplication—in conjunction with prime factorization—to reduce common fractions. The Identity Property of Multiplication is true read from left to right or from right to left. If

$$\frac{2}{3} = \frac{2}{3} \cdot \frac{2}{2} = \frac{4}{6}$$

then

$$\frac{4}{6} = \frac{2}{3} \cdot \frac{2}{2} = \frac{2}{3}$$

EXAMPLE 5 Using the prime factorization method, reduce $\frac{6}{18}$. To begin, write the given fraction and its prime factored form:

To *reduce a common fraction,* we convert a given fraction to an equivalent that has a smaller denominator using the following steps:

1. Write the given common fraction.

2. Express the numerator and denominator in prime factored form.

3. Reduce the fraction by giving those factors common to both numerator and denominator the equivalent name of 1 $\left(\frac{2}{2}=1,\right.$ $\frac{3}{3}=1,$ and so on$\Big)$.

4. The product of those factors left in the numerator is the numerator of the reduced fraction. The product of those factors left in the denominator is the denominator of the reduced fraction.

EXAMPLE 5
continued

$$\frac{6}{18}=\frac{2\cdot3}{2\cdot3\cdot3} \tag{1, 2}$$

In this instance there are two pairs of common factors, $\frac{2}{2}$ and $\frac{3}{3}$.

$$=\frac{\cancel{2}^{1}\cdot\cancel{3}^{1}}{\cancel{2}\cdot\cancel{3}\cdot3} \tag{3}$$

Now, use the product of those factors left to produce the reduced fraction:

$$=\frac{1}{3} \tag{4}$$

EXAMPLE 6

Reduce $\frac{14}{18}$.

$$\frac{14}{18}=\frac{2\cdot7}{2\cdot3\cdot3} \tag{1, 2}$$

$$=\frac{\cancel{2}\cdot7}{\cancel{2}\cdot3\cdot3} \tag{3}$$

$$=\frac{7}{9} \tag{4}$$

EXAMPLE 7 Reduce $\frac{14}{15}$.

$$\frac{14}{15} = \frac{2 \cdot 7}{3 \cdot 5}$$

$$= \frac{14}{15}$$

In Example 7 there is no factor common to both numerator and denominator, so $\frac{14}{15}$ is reduced as far as it can be—to its lowest terms. When a common fraction is reduced to its *lowest terms*, there are no factors common to both numerator and denominator.

Exercises 1.3 Build equivalent fractions using the given denominator.

1. $\dfrac{1}{2} = \dfrac{}{6}$ 2. $\dfrac{1}{3} = \dfrac{}{6}$

3. $4 = \dfrac{}{20}$ 4. $7 = \dfrac{}{3}$

5. $2 = \dfrac{}{5}$ 6. $\dfrac{1}{8} = \dfrac{}{24}$

7. $\dfrac{3}{16} = \dfrac{}{336}$ 8. $\dfrac{7}{20} = \dfrac{}{60}$

9. $\dfrac{1}{16} = \dfrac{}{48}$ 10. $\dfrac{7}{9} = \dfrac{}{108}$

Reduce by the prime factorization method.

11. $\dfrac{15}{35}$ 12. $\dfrac{16}{18}$

13. $\dfrac{100}{105}$ 14. $\dfrac{24}{62}$

15. $\dfrac{421}{483}$ 16. $\dfrac{4914}{13\,860}$

17. $\dfrac{42}{60}$ 18. $\dfrac{54}{66}$

19. $\dfrac{9}{87}$ 20. $\dfrac{437}{667}$

Least Common Multiples   Section 1.4

> To add and subtract common fractions, it is necessary not only to build equivalent fractions (Section 1.3) but also to find the least common multiple of a set of numbers. We use the following procedure to find the least common multiple (lcm) of two or more numbers:
>
> 1. List the numbers.
> 2. List the prime factorization of each number.
> 3. Construct the least common multiple by listing each factor that appears in any given number. List each factor the greatest number of times it appears in any one number.

Find the least common multiple of 12 and 15. To begin, list the numbers and the prime factorization of each. Then construct the lcm.

EXAMPLE 1

numbers		prime factorizations	least common multiple
12	=	$2 \cdot 2 \cdot 3$	$2 \cdot 2 \cdot 3 \cdot 5 = 60$
15	=	$3 \cdot 5$	

The lcm of 12 and 15 is 60 and 60 is the smallest number divisible by both 12 and 15.

Looking at Example 1, note that we repeat the 2 in the least common multiple ($2 \cdot 2 \cdot 3 \cdot 5$ or 60) because the factor 2 appears twice in the prime factorization of 12 ($2 \cdot 2 \cdot 3$). We list the factors 3 and 5 just once in the lcm ($2 \cdot 2 \cdot 3 \cdot 5$) because each appears just once in each of the prime factorizations. That is, 3 appears once in $2 \cdot 2 \cdot 3$ (12) and once in $3 \cdot 5$ (15); 5 appears once in $3 \cdot 5$ (15). The lcm (60) is a multiple of 12 ($60 = 12 \cdot 5$) and a multiple of 15 ($60 = 15 \cdot 4$).

EXAMPLE 2

Find the least common multiple of 15, 35, and 42.

numbers		prime factorizations	least common multiple
15	=	$3 \cdot 5$	
35	=	$5 \cdot 7$	$2 \cdot 3 \cdot 5 \cdot 7 = 210$
42	=	$2 \cdot 3 \cdot 7$	

Since each factor (2, 3, 5, and 7) appears only one time in any one number, each factor is listed only one time in the lcm.

EXAMPLE 3

Find the lcm of 12, 15, and 18.

numbers		prime factorizations	least common multiple
12	=	$2 \cdot 2 \cdot 3$	
15	=	$3 \cdot 5$	$2 \cdot 2 \cdot 3 \cdot 3 \cdot 5 = 180$
18	=	$2 \cdot 3 \cdot 3$	

The factor 2 appears twice in 12 ($2 \cdot 2 \cdot 3$) and is listed twice in the lcm. The factor 3 appears twice in 18 ($2 \cdot 3 \cdot 3$) and is listed twice in the lcm. The factor 5 appears one time in any one number and is listed only one time in the lcm.

EXAMPLE 4

Find the lcm of 10 and 21.

numbers		prime factorizations	least common multiple
10	=	$2 \cdot 5$	$2 \cdot 5 \cdot 3 \cdot 7 = 210$
21	=	$3 \cdot 7$	

Since there are no common factors in the prime factorizations of 10 and 21, their least common multiple is 10 times 21 or 210.

Exercises 1.4

In Exercises 1–16 find the least common multiple of each set of numbers. Show your work.

1. 2 and 6
2. 6 and 8
3. 12 and 15
4. 10 and 21
5. 24 and 40
6. 16 and 42
7. 2, 3, and 6
8. 3, 8, and 12
9. 6, 10, and 45
10. 12, 16, and 24
11. 14 and 15
12. 36, 56, and 72

13. 3, 6, and 20   14. 21, 24, and 28

15. 4, 8, 16, and 24   16. 26, 33, and 35

Addition and Subtraction of Common Fractions

If the fractions' denominators are the same, add or subtract the numerators and use as a denominator the common denominator.

$$\frac{7}{11} + \frac{2}{11} = \frac{9}{11}$$

$$\frac{7}{11} - \frac{2}{11} = \frac{5}{11}$$

To solve a problem with unlike denominators $\left(\frac{1}{2}\right.$ plus $\frac{1}{3}$, for instance$\left.\right)$ is somewhat more complicated. It is necessary to find fractions equivalent to $\frac{1}{2}$ and $\frac{1}{3}$ that have the same denominator. We call the needed denominator the *least common denominator* (lcd), or least common multiple, of all the denominators in the problem.

To add or subtract follow this procedure:

1. Find the least common denominator (lcd) by finding the least common multiple (lcm) of the denominators (Section 1.4).

2. Build fractions equivalent to the given fractions using as a denominator the least common denominator (Section 1.3).

3. Add or subtract the numerators and use as a denominator the common denominator.

4. Reduce the answer if necessary (Section 1.3).

Using the procedure, add $\frac{1}{2}$ and $\frac{1}{3}$. To begin, find the least common denominator of 2 and 3, which is 6. Now, build fractions equivalent to $\frac{1}{2}$ and $\frac{1}{3}$ using the lcd (6) as the denominator:

EXAMPLE 1

$$\frac{1}{2} + \frac{1}{3} = \left(\frac{1}{2} \cdot \frac{3}{3}\right) + \left(\frac{1}{3} \cdot \frac{2}{2}\right) \qquad (1, 2)$$

EXAMPLE 1
continued

$$= \frac{3}{6} + \frac{2}{6}$$

Add the numerators, keeping the lcd as the denominator:

$$= \frac{5}{6} \tag{3}$$

EXAMPLE 2

$$\frac{5}{18} + \frac{1}{12} - \frac{2}{15} =$$

To begin, find the lcm of 12, 15, and 18 (Section 1.4, Example 3). Using the lcm (180) as the least common denominator, build equivalent fractions (Section 1.3, Examples 2–4):

$$= \left(\frac{5}{18} \cdot \frac{10}{10} \right) + \left(\frac{1}{12} \cdot \frac{15}{15} \right) - \left(\frac{2}{15} \cdot \frac{12}{12} \right)$$

$$= \frac{50}{180} + \frac{15}{180} - \frac{24}{180}$$

The denominators are now the same, so add or subtract as indicated:

$$= \frac{41}{180}$$

In Examples 1 and 2 the answers are in lowest terms. In Example 3, reduction is necessary.

EXAMPLE 3

$$\frac{2}{15} - \frac{1}{42} + \frac{7}{35} =$$

To begin, find the lcm of 15, 35, and 42 (Section 1.4, Example 2). Using the lcm (210) as the least common denominator, build equivalent fractions and then add or subtract as indicated:

$$= \left(\frac{2}{15} \cdot \frac{14}{14} \right) - \left(\frac{1}{42} \cdot \frac{5}{5} \right) + \left(\frac{7}{35} \cdot \frac{6}{6} \right)$$

$$= \frac{28}{210} - \frac{5}{210} + \frac{42}{210}$$

$$= \frac{65}{210}$$

Now reduce the answer:

EXAMPLE 3
continued

$$\frac{65}{210} = \frac{5 \cdot 13}{2 \cdot 3 \cdot 5 \cdot 7}$$

$$= \frac{\cancel{5} \cdot 13}{2 \cdot 3 \cdot \cancel{5} \cdot 7}$$

$$= \frac{13}{42}$$

Add or subtract as indicated. Show your work.

Exercises 1.5

1. $\frac{1}{6} + \frac{1}{8} =$

2. $\frac{1}{24} + \frac{1}{40} =$

3. $\frac{3}{16} - \frac{5}{42} =$

4. $\frac{1}{12} + \frac{1}{16} - \frac{1}{24} =$

5. $\frac{1}{36} + \frac{1}{56} + \frac{1}{72} =$

6. $\frac{5}{12} - \frac{2}{12} + \frac{5}{18} =$

7. $\frac{2}{3} + \frac{5}{6} + \frac{7}{20} =$

8. $\frac{1}{8} + \frac{1}{4} + \frac{1}{16} + \frac{1}{24} =$

9. $\frac{6}{35} - \frac{2}{63} + \frac{7}{45} =$

10. $\frac{1}{100} - \frac{1}{150} =$

Improper Fractions and Mixed Numbers

Section 1.6

To this point we've been discussing *proper fractions*—fractions that have numerators smaller than their denominators. Fractions that have numerators larger than their denominators are called *improper fractions*. $\frac{5}{3}$ and $\frac{21}{5}$ are improper fractions. By dividing, we can change improper fractions to whole numbers or to *mixed numbers*—numbers that have a whole number part and a fraction part—for example, $4\frac{1}{2}$.

Change $\frac{6}{3}$ to a whole number.

EXAMPLE 1

$$\frac{6}{3} = 6 \div 3$$

$$= 3\overline{)6}$$

$$= 2$$

EXAMPLE 2   Change $\frac{21}{5}$ to a mixed number.

$$\frac{21}{5} = 21 \div 5$$

$$= 5\overline{)21}$$

$$= 4\frac{1}{5}$$

Mixed numbers may be changed to improper fractions using this formula:

mixed number

$$\text{whole number} \frac{\text{numerator}}{\text{denominator}}$$

improper fraction

$$= \frac{(\text{denominator}) \cdot (\text{whole number}) + \text{numerator}}{\text{denominator}}$$

EXAMPLE 3   Convert $6\frac{3}{8}$ to an improper fraction.

$$6\frac{3}{8} = \frac{(8 \cdot 6) + 3}{8}$$

$$= \frac{48 + 3}{8}$$

$$= \frac{51}{8}$$

EXAMPLE 4   Convert $4\frac{1}{5}$ to an improper fraction.

$$4\frac{1}{5} = \frac{(5 \cdot 4) + 1}{5}$$

$$= \frac{20 + 1}{5}$$

$$= \frac{21}{5}$$

To add mixed numbers, we can express them in addition form or change them to improper fractions. Both methods are illustrated here.

Add $2\frac{1}{4}$ and $5\frac{2}{3}$. Using the addition form, separate the whole numbers from their fraction parts:

EXAMPLE 5

$$2\frac{1}{4} + 5\frac{2}{3} = 2 + \frac{1}{4} + 5 + \frac{2}{3}$$

Now add the whole numbers:

$$= 2 + 5 + \frac{1}{4} + \frac{2}{3}$$

$$= 7 + \frac{1}{4} + \frac{2}{3}$$

Finally, to add the fraction determine the lcd (12), build fractions equivalent to $\frac{1}{4}$ and $\frac{2}{3}$ using 12 as the denominator, and then carry out the addition.

$$= 7 + \left(\frac{1}{4} \cdot \frac{3}{3}\right) + \left(\frac{2}{3} \cdot \frac{4}{4}\right)$$

$$= 7 + \frac{3}{12} + \frac{8}{12}$$

$$= 7 + \frac{11}{12}$$

$$= 7\frac{11}{12}$$

Now, let's work the same problem using the second method—converting to improper fractions.

$$2\frac{1}{4} + 5\frac{2}{3} = \frac{9}{4} + \frac{17}{3}$$

EXAMPLE 6

Once again, find the lcd and build equivalent fractions:

$$= \left(\frac{3}{3} \cdot \frac{9}{4}\right) + \left(\frac{17}{3} \cdot \frac{4}{4}\right)$$

$$= \frac{27}{12} + \frac{68}{12}$$

EXAMPLE 6
continued

You can see that with this method we work with larger numbers. Now, finish the addition, reducing the answer:

$$= \frac{95}{12}$$

$$= 7\frac{11}{12}$$

To subtract mixed numbers you can avoid borrowing or re-grouping by changing the mixed numbers to improper fractions and then subtracting.

EXAMPLE 7

$$5\frac{1}{4} - 2\frac{3}{8} = \frac{21}{4} - \frac{19}{8}$$

$$= \left(\frac{2}{2} \cdot \frac{21}{4}\right) - \frac{19}{8}$$

$$= \frac{42}{8} - \frac{19}{8}$$

$$= \frac{23}{8}$$

$$= 2\frac{7}{8}$$

Exercises 1.6   Convert to mixed numbers.

1. $\frac{15}{4}$

2. $\frac{63}{15}$

3. $\frac{17}{2}$

4. $\frac{24}{7}$

5. $\frac{45}{22}$

6. $\frac{102}{72}$

7. $\frac{5}{3}$

8. $\frac{37}{8}$

Convert to improper fractions.

9. $8\frac{1}{5}$

10. $4\frac{3}{8}$

11. $7\frac{2}{9}$

12. $2\frac{1}{150}$

13. $3\dfrac{1}{3}$ 14. $33\dfrac{1}{3}$

15. $16\dfrac{2}{3}$ 16. $11\dfrac{7}{32}$

In Exercises 17–26 add or subtract as indicated. Show your work.

17. $3 - 2\dfrac{1}{2} =$ 18. $3\dfrac{1}{3} - 2\dfrac{1}{2} =$

19. $6\dfrac{1}{5} + 5\dfrac{2}{19} =$ 20. $3 + 5\dfrac{6}{17} + 4\dfrac{2}{3} =$

21. $5 + 7\dfrac{3}{8} - 3\dfrac{1}{4} =$ 22. $45 - 7\dfrac{5}{13} =$

23. $7\dfrac{3}{8} - 3\dfrac{1}{4} =$ 24. $6\dfrac{2}{3} + 3 - 2\dfrac{1}{6} =$

25. $5\dfrac{1}{25} - 4\dfrac{1}{15} =$ 26. $2\dfrac{1}{30} - 1\dfrac{1}{40} + \dfrac{1}{200} =$

27. An infant has ingested $3\frac{1}{2}$ ounces milk, $1\frac{3}{4}$ ounces water, $3\frac{2}{3}$ ounces milk, 4 ounces milk, $1\frac{1}{8}$ ounces water, and $2\frac{3}{4}$ ounces milk. What is the infant's total liquid intake?

Multiplication and Division Involving Common Fractions

Section 1.7

To multiply common fractions, use the same solution process used to reduce fractions (Section 1.3). In the reduction of common fractions we deal with one numerator and one denominator—a single fraction; in the multiplication of common fractions we deal with two or more numerators and denominators—two or more fractions.

Multiplication

$$\frac{21}{35} \cdot \frac{2}{5} \cdot \frac{5}{33} =$$

$$= \frac{3 \cdot 7}{5 \cdot 7} \cdot \frac{2}{5} \cdot \frac{5}{3 \cdot 11}$$

(1, 2)

EXAMPLE 1

To multiply common fractions or to reduce common fractions:

1. Write the problem.
2. Express each numerator and denominator in prime factored form.
3. Reduce.
4. Multiply those factors left in the numerators; multiply those factors left in the denominators.

EXAMPLE 1
continued

$$= \frac{\cancel{3}^1 \cdot \cancel{7}^1}{5 \cdot \cancel{7}} \cdot \frac{2}{\cancel{3}} \cdot \frac{\cancel{3}^1}{\cancel{3} \cdot 11} \quad * \tag{3}$$

$$= \frac{2}{5 \cdot 11} \tag{4}$$

$$= \frac{2}{55}$$

If a multiplication problem contains mixed numbers, convert them to improper fractions (Section 1.6). Also, convert whole numbers to fraction form: 2 becomes $\frac{2}{1}$; 15 becomes $\frac{15}{1}$.

EXAMPLE 2

$$2\frac{1}{5} \cdot 3\frac{3}{4} \cdot 2 = \frac{11}{5} \cdot \frac{15}{4} \cdot \frac{2}{1} \tag{1, 2}$$

$$= \frac{11}{5} \cdot \frac{3 \cdot 5}{2 \cdot 2} \cdot \frac{2}{1}$$

$$= \frac{11}{\cancel{5}} \cdot \frac{3 \cdot \cancel{5}^1}{2 \cdot \cancel{2}} \cdot \frac{\cancel{2}^1}{1} \tag{3}$$

$$= \frac{11 \cdot 3}{2 \cdot 1} \tag{4}$$

$$= \frac{33}{2}$$

$$= 16\frac{1}{2}$$

---

*Each pair of factors common to numerator and denominator is given the equivalent name of 1. These pairs may be directly over each other $\left(\frac{7}{7}\right)$ or separated $\left(\frac{3}{3} \text{ and } \frac{5}{5}\right)$.

Note that the final answer can be written as an improper fraction or as a mixed number, whichever is needed.

In the solution of medication problems, the word *of* frequently indicates multiplication.

Verbal Expressions

You are asked to administer 10 millilitres solution. You have given two-fifths *of* the medication; how much have you given? To begin, look for the verbal clue:

EXAMPLE 3

$$\text{two-fifths } of \text{ the medication} = \frac{2}{5} \cdot \frac{10}{1} \text{ millilitres}$$

$$= \frac{2}{5} \cdot \frac{2 \cdot 5}{1}$$

$$= \frac{2}{\cancel{5}} \cdot \frac{2 \cdot \cancel{5}}{1}$$

$$= \frac{4}{1}$$

$$= 4 \text{ millilitres have been given}$$

Here is another type of problem that requires multiplication.

Give two and one-half tablets three times daily. How many tablets are needed for one week? To begin, translate the problem to arithmetic:

EXAMPLE 4

two and one-half tablets, three times daily, one week

$$= 2\frac{1}{2} \cdot 3 \cdot 7$$

Then proceed with the multiplication:

$$= \frac{5}{2} \cdot \frac{3}{1} \cdot \frac{7}{1}$$

$$= \frac{105}{2}$$

$$= 52\frac{1}{2} \text{ tablets needed for one week}$$

To divide common fractions, invert the divisor and multiply.

Division

EXAMPLE 5

$$\frac{6}{35} \div \frac{12}{25} \text{ (divisor)} = \frac{6}{35} \cdot \frac{25}{12}$$

$$= \frac{2 \cdot 3}{5 \cdot 7} \cdot \frac{5 \cdot 5}{2 \cdot 2 \cdot 3}$$

$$= \frac{\cancel{2}^1 \cdot \cancel{3}^1}{\cancel{5} \cdot 7} \cdot \frac{5 \cdot \cancel{5}^1}{\cancel{2} \cdot 2 \cdot \cancel{3}}$$

$$= \frac{5}{7 \cdot 2}$$

$$= \frac{5}{14}$$

When mixed numbers are involved, change them to improper fractions and then divide.

EXAMPLE 6

$$2\frac{1}{2} \div 1\frac{1}{6} = \frac{5}{2} \div \frac{7}{6}$$

$$= \frac{5}{2} \cdot \frac{6}{7}$$

$$= \frac{5}{\cancel{2}} \cdot \frac{3 \cdot \cancel{2}}{7}$$

$$= \frac{5 \cdot 3}{7}$$

$$= \frac{15}{7}$$

$$= 2\frac{1}{7}$$

Verbal Expressions

Verbal expressions that indicate division. The verbal statements *is what part of* and *is how many times* each indicate the operation of division. In fact, the division symbol may be substituted for both of these phrases.

EXAMPLE 7

One-third ounce is what part of two-thirds ounce?

$$\text{one-third } is \ what \ part \ of \text{ two-thirds} = \frac{1}{3} \div \frac{2}{3}$$

EXAMPLE 7
continued

$$= \frac{1}{3} \cdot \frac{3}{2}$$

$$= \frac{1}{\cancel{3}} \cdot \frac{\cancel{3}}{2}$$

$$= \frac{1}{2}$$

Or, one-third ounce is one-half of two-thirds ounce.

Two and one-half is how many times one-half?

EXAMPLE 8

$$\text{two and one-half } \textit{is how many times one half} = 2\frac{1}{2} \div \frac{1}{2}$$

$$= \frac{5}{2} \div \frac{1}{2}$$

$$= \frac{5}{2} \cdot \frac{2}{1}$$

$$= \frac{5}{\cancel{2}} \cdot \frac{\cancel{2}}{1}$$

$$= \frac{5}{1}$$

$$= 5$$

Or, two and one-half is five times one-half.

In Exercises 1–20 multiply or divide as indicated. Show your work.   **Exercises 1.7**

1. $\dfrac{1}{4} \cdot \dfrac{1}{2} =$

2. $\dfrac{3}{4} \cdot \dfrac{10}{25} \cdot \dfrac{2}{21} =$

3. $\dfrac{1}{14} \cdot \dfrac{15}{33} \cdot \dfrac{42}{195} =$

4. $1\dfrac{1}{2} \cdot \dfrac{5}{6} =$

5. $1\dfrac{1}{2} \cdot 2\dfrac{1}{4} =$

6. $\dfrac{1}{100} \cdot 3\dfrac{1}{3} =$

7. $\dfrac{1}{100} \cdot 4\dfrac{5}{8} \cdot 100 =$

8. $1\dfrac{2}{13} \cdot 7\dfrac{2}{9} \cdot 2\dfrac{2}{5} =$

9. $2\dfrac{1}{3} \cdot 3\dfrac{1}{7} \cdot \dfrac{4}{11} =$

10. $\dfrac{16}{5} \cdot 3\dfrac{1}{2} \cdot 4 =$

11. $\dfrac{1}{150} \div \dfrac{3}{4} =$ 12. $\dfrac{3}{4} \div \dfrac{1}{150} =$

13. $3 \div 4\dfrac{3}{8} =$ 14. $4\dfrac{3}{8} \div 3 =$

15. $\dfrac{9}{10} \div 100 =$ 16. $\dfrac{9}{10} \div \dfrac{1}{100} =$

17. $3\dfrac{7}{9} \div 50 =$ 18. $3\dfrac{5}{8} \div 4\dfrac{7}{32} =$

19. $\dfrac{34}{12} \div \dfrac{10}{13} =$ 20. $10\dfrac{2}{3} \cdot \dfrac{17}{25} \cdot \dfrac{21}{14} \div \dfrac{64}{15} =$

21. The doctor asks you to administer $7\frac{1}{2}$ tablets daily. You have given half of them; how many tablets have you given?

22. The doctor asks you to administer $7\frac{1}{2}$ tablets daily. If each dose is $1\frac{1}{2}$ tablets, how many times a day must you administer the tablets?

23. $3\frac{3}{4}$ is how many times $1\frac{3}{4}$?

24. Two and three-fourths ounces is what part of 11 ounces?

25. What is $\frac{5}{6}$ of $4\frac{4}{5}$ ounces?

26. Administer $1\frac{1}{2}$ ounces drug A three times daily and $\frac{1}{6}$ ounce drug B twice daily. What is the patient's total daily drug consumption?

27. $\frac{1}{50}$ is how many times $\frac{1}{100}$?

28. $\frac{1}{300}$ is what part of $\frac{1}{150}$?

Section 1.8   Complex Fractions

A *complex fraction* is a fraction in which the numerator or denominator (or both) is a fraction, like the following:.

$$\dfrac{\dfrac{1}{2}}{\dfrac{3}{5}} \qquad \dfrac{6}{\dfrac{1}{7}} \qquad \dfrac{\dfrac{1}{5}+\dfrac{2}{3}}{5} \qquad \dfrac{150}{2\dfrac{1}{5}}$$

Note that the main fraction bar is longer than the others.

We can express complex fractions in a simpler form by performing the division implied in the fraction:

EXAMPLE 1

$$\frac{\frac{1}{2}}{\frac{3}{5}} = \frac{1}{2} \div \frac{3}{5}$$

$$= \frac{1}{2} \cdot \frac{5}{3}$$

$$= \frac{5}{6}$$

EXAMPLE 2

$$\frac{6}{\frac{1}{7}} = \frac{6}{1} \div \frac{1}{7}$$

$$= \frac{6}{1} \cdot \frac{7}{1}$$

$$= 42$$

EXAMPLE 3

$$\frac{\frac{1}{5} + \frac{2}{3}}{5} = \left(\frac{1}{5} + \frac{2}{3}\right) \div \frac{5}{1}$$

$$= \left(\frac{3}{15} + \frac{10}{15}\right) \div \frac{5}{1}$$

$$= \frac{13}{15} \div \frac{5}{1}$$

$$= \frac{13}{15} \cdot \frac{1}{5}$$

$$= \frac{13}{75}$$

EXAMPLE 4

$$\frac{150}{2\frac{1}{5}} = \frac{150}{1} \div 2\frac{1}{5}$$

EXAMPLE 4
continued

$$= \frac{150}{1} \div \frac{11}{5}$$

$$= \frac{150}{1} \cdot \frac{5}{11}$$

$$= \frac{750}{11}$$

$$= 68\frac{2}{11}$$

Exercises 1.8   Simplify the following complex fractions.

1. $\dfrac{\dfrac{1}{150}}{\dfrac{1}{75}}$

2. $\dfrac{\dfrac{1}{9}}{\dfrac{1}{3}}$

3. $\dfrac{7\dfrac{1}{2}}{15}$

4. $\dfrac{6}{\dfrac{1}{10}}$

5. $\dfrac{9\dfrac{1}{8}}{4\dfrac{3}{8}}$

6. $\dfrac{\dfrac{1}{200}}{\dfrac{3}{150}}$

7. $\dfrac{3}{\dfrac{3}{8}}$

8. $\dfrac{\dfrac{3}{8}}{3}$

9. $\dfrac{\dfrac{1}{100}}{\dfrac{1}{4}}$

10. $\dfrac{\dfrac{1}{4}+\dfrac{1}{8}}{\dfrac{1}{2}+\dfrac{1}{4}}$

Comparing Fractions

When we compare the size of any two quantities, the first may be equal to, greater than, or less than the second. We use the following symbols to indicate these relationships:

equal to $=$ $\left(\dfrac{1}{2}=\dfrac{3}{6}\right)$

greater than $>$ $\left(\dfrac{2}{3}>\dfrac{1}{3}\right)$

less than $<$ $\left(\dfrac{1}{4}<\dfrac{3}{4}\right)$

And there is a fourth symbol that we also use frequently:

does not equal $\neq$ $\left(\dfrac{1}{2}\neq\dfrac{3}{5}\right)$

If two fractions are equal, their cross products also are equal. By the same token, if the cross products of two fractions are equal, then the fractions also are equal. That is, if

Cross Products and Equality

$$\frac{1}{2}=\frac{3}{6}$$

then

$$6\cdot 1 = 2\cdot 3$$

and if

$$6\cdot 1 = 2\cdot 3$$

then

$$\frac{1}{2}=\frac{3}{6}$$

if

$$\frac{2}{3} = \frac{6}{9}$$

then

$$3 \cdot 6 = 9 \cdot 2 \ (18 = 18)$$

and if

$$3 \cdot 6 = 9 \cdot 2$$

then

$$\frac{2}{3} = \frac{6}{9}$$

We can use the cross product check to determine equality.

EXAMPLE 1 Does $\frac{3}{4}$ equal $\frac{15}{20}$?

$$\frac{3}{4} \quad \frac{15}{20}$$

$$20 \cdot 3 = 4 \cdot 15$$

$$60 = 60$$

$$\frac{3}{4} = \frac{15}{20}$$

EXAMPLE 2 Does $\frac{5}{6}$ equal $\frac{8}{9}$?

$$\frac{5}{6} \quad \frac{8}{9}$$

$$9 \cdot 5 \neq 6 \cdot 8$$

$$45 \neq 48$$

$$\frac{5}{6} \neq \frac{8}{9}$$

A variation of the cross product check allows us to determine which unequal fraction is greater. To begin, simply write the cross product near the numerator involved in the multiplication. The side with the greater product will be the greater fraction.

Which is greater, $\frac{2}{3}$ or $\frac{1}{3}$?

$$\frac{2}{3} \quad \frac{1}{3}$$

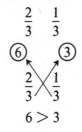

$$6 > 3$$

therefore,

$$\frac{2}{3} > \frac{1}{3}$$

EXAMPLE 3

Which is smaller, $\frac{1}{4}$ or $\frac{3}{4}$?

$$\frac{1}{4} \quad \frac{3}{4}$$

④ ⑫

$$\frac{1}{4} \diagup\!\!\!\!\diagdown \frac{3}{4}$$

$$4 < 12$$

therefore,

$$\frac{1}{4} < \frac{3}{4}$$

EXAMPLE 4

EXAMPLE 5 | Which is greater, $\frac{1}{100}$ or $\frac{1}{200}$?

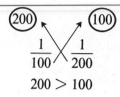

$$200 > 100$$

therefore,

$$\frac{1}{100} > \frac{1}{200}$$

EXAMPLE 6 | Which is greater, $\frac{3}{5}$ or $\frac{4}{7}$?

$$\frac{3}{5} \qquad \frac{4}{7}$$

$$21 > 20$$

therefore,

$$\frac{3}{5} > \frac{4}{7}$$

Exercises 1.9 In Exercises 1–10 make each statement true by placing $=$, $>$, or $<$ between each pair of numbers.

1. $\frac{1}{3}$ $\frac{1}{2}$ 2. $\frac{2}{5}$ $\frac{1}{4}$

3. $\frac{3}{8}$ $\frac{3}{11}$ 4. $\frac{1}{200}$ $\frac{1}{150}$

5. $\frac{1}{120}$ $\frac{1}{125}$ 6. $\frac{6}{13}$ $\frac{5}{11}$

7. $\frac{5}{8}$ $\frac{7}{11}$ 8. $\frac{6}{9}$ $\frac{4}{6}$

9. $\frac{4}{7}$ $\frac{2}{3}$ 10. $\frac{7}{18}$ $\frac{5}{12}$

In Exercises 11–14 put the fractions into order from smallest to largest.

11. $\dfrac{7}{18}, \dfrac{1}{3}, \dfrac{5}{12}$

12. $\dfrac{2}{3}, \dfrac{5}{8}, \dfrac{4}{7}$

13. $\dfrac{5}{8}, \dfrac{7}{11}, \dfrac{6}{13}$

14. $\dfrac{4}{5}, \dfrac{3}{4}, \dfrac{9}{10}, \dfrac{1}{2}, \dfrac{7}{8}, \dfrac{2}{3}, \dfrac{5}{6}$

15. Is $\frac{1}{3}$ more nearly $\frac{3}{10}$ or $\frac{33}{100}$?

[1.1]

1. Using one tablet, illustrate each of the following common fractions:

    a. $\dfrac{1}{4}$ b. $\dfrac{2}{3}$

    c. $\dfrac{1}{2}$ d. $\dfrac{3}{4}$

2. You began with 18 millilitres liquid and administered 5 millilitres to a patient. Express as a common fraction

    a. the part given.   b. the part remaining.

[1.2]

3. State whether each number is divisible by 2, by 3, or by 5.

    a. 85   b. 169

    c. 195   d. 378

4. Find the prime factorization of each number. Show your work.

    a. 51   b. 169

    c. 114   d. 200

[1.3]

5. Using the given denominator, build equivalent fractions.

    a. $\dfrac{3}{13} = \dfrac{}{65}$ b. $\dfrac{1}{8} = \dfrac{}{160}$

c. $6 = \dfrac{}{5}$ d. $\dfrac{7}{5} = \dfrac{}{35}$

6. Using the prime factorization method, reduce the following fractions.

     a. $\dfrac{135}{1430}$ b. $\dfrac{57}{78}$

     c. $\dfrac{38}{899}$ d. $\dfrac{45}{150}$

[1.4]

7. Find the lcm of each set of numbers.

     a.   9, 15, and 21   b.   4, 12, and 18

     c.   6, 11, and 35   d.   6, 35, 42

[1.6]

8. Change these improper fractions to mixed numbers.

     a. $\dfrac{23}{4}$ b. $\dfrac{17}{9}$

     c. $\dfrac{38}{8}$ d. $\dfrac{15}{4}$

9. Change these mixed numbers to improper fractions.

     a. $7\dfrac{1}{8}$ b. $4\dfrac{3}{4}$

     c. $10\dfrac{5}{8}$ d. $4\dfrac{5}{6}$

10. Add or subtract as indicated.

     a. $\dfrac{5}{21} + \dfrac{7}{15} + \dfrac{8}{9} =$ b. $\dfrac{1}{150} - \dfrac{1}{200} =$

     c. $7 + 2\dfrac{1}{8} - 5\dfrac{7}{9} =$ d. $3\dfrac{1}{2} + \dfrac{1}{10} + 5 =$

[1.7]

11. Multiply or divide as indicated.

     a. $\dfrac{1}{150} \div \dfrac{1}{200} =$ b. $\dfrac{3}{7} \cdot \dfrac{51}{108} \cdot \dfrac{91}{153} =$

     c. $2\dfrac{8}{15} \cdot \dfrac{7}{8} \div 1\dfrac{2}{5} =$ d. $2\dfrac{1}{2} \cdot \dfrac{1}{150} =$

12. $5\frac{3}{4}$ is how many times $1\frac{1}{2}$?

13. $\frac{1}{200}$ is what part of $\frac{1}{150}$?

14. What is $\frac{7}{8}$ of $8\frac{1}{2}$?

[1.8]

15. Simplify these complex fractions.

a. $\dfrac{\dfrac{3}{4}}{\dfrac{5}{8}}$

b. $\dfrac{2\dfrac{3}{4}}{\dfrac{15}{16}}$

c. $\dfrac{\dfrac{1}{150}}{\dfrac{1}{200}}$

d. $\dfrac{\dfrac{1}{150}}{3}$

[1.9]

16. Make each statement true by placing $=$, $>$, or $<$ between each pair of numbers.

a. $\dfrac{1}{300}$ $\dfrac{1}{100}$

b. $\dfrac{7}{9}$ $\dfrac{10}{13}$

c. $\dfrac{13}{17}$ $\dfrac{52}{68}$

d. $\dfrac{1}{100}$ $\dfrac{1}{200}$

Decimal Numerals

With the advent of the hand calculator, the computer, the measurement of drugs in metric units and the labeling of equipment in metric units, the use of the decimal fraction in health-related problems increased. Increased use means increased chance of error. For example, when medication is being administered, an error of one decimal position means either ten times too much medicine or just one-tenth of the prescribed dosage. Either case is unacceptable, so in this chapter careful attention is paid to working forms that decrease chance of error.

Topics include:

Reading and writing decimal numerals

Working with powers of ten (10)

Adding, subtracting, multiplying, and dividing with decimal numerals

Converting decimal numerals to common fractions and vice versa

Comparing sizes of decimal numerals

[2.1]
  1. Write 1000 as a power of 10 (exponential form).

[2.2]
  2. Write $\frac{1}{10^3}$ as a decimal numeral.
  3. Write *forty and ninety ten-thousandths* as a decimal numeral.
  4. Write 10.0105 in words.
  5. Write $\frac{14}{1\,000\,000}$ as a decimal numeral.

[2.3]
  6. $2.3 + 0.09 + 13 =$
  7. $9 - 0.059 =$

[2.4]
  8. $0.0025 \times 1.04 =$
  9. $3.007 \times 10^2 =$
 10. What is 0.02 of 24.5?

[2.5]
 11. $0.267 \div 0.003 =$
 12. $18 \div 1.02 =$ (correct to the hundredth place)
 13. $0.209 \div 3 =$ (correct to the thousandth place)
 14. $1.25 \div 1000 =$
 15. 0.09 is what part of 8.1?
 16. 0.6 is how many times 0.003?

[2.6]
 17. Place $=$, $>$, or $<$ between each pair of numbers to make a
      true statement.

      a. 0.0099   0.010   b. $\frac{2}{33}$ 0.06

      c. $\frac{1}{8}$ 0.0125   d. 2.05 $2\frac{50}{100}$

[2.7]
 18. Change 0.055 to an equivalent common fraction.
 19. Change $\frac{1}{150}$ to an equivalent decimal numeral.
 20. Change 6.03 to an equivalent common fraction.

Section 2.1 **Exponents**

Work with the decimal system of numeration is made easier by some understanding of numbers expressed in product form, in factored form, and in exponential form. For example,

when 8 is written *8,* it is in *product form*

when 8 is written *2 · 2 · 2,* it is in *factored form*

when 8 is written as *2³,* it is in *exponential form*

In Chapter 1 the factor and product forms were discussed; the exponential form will be discussed here.

Look again at 2^3 (8 in exponential form). The 2 is called the *base,* and the ^{3} is called the *exponent* or *power.* 2^3 is read *two cubed* or *two to the third power.* The exponent or power (^{3}) tells how many times to write the base (2) as a factor.

EXAMPLE 1					

exponential form		factored form		product form
2^3	$=$	$2 \cdot 2 \cdot 2$	$=$	8

The base 2 is written as a factor three times because 3 is the exponent or power.

EXAMPLE 2

$$3^2 = 3 \cdot 3 = 9$$

The base 3 is written as a factor two times because the exponent is 2.

EXAMPLE 3

$$3^4 = 3 \cdot 3 \cdot 3 \cdot 3 = 81$$

The base 3 is written as a factor four times because the exponent is 4.

EXAMPLE 4

The converse is also true. If

$$2^3 = 2 \cdot 2 \cdot 2 = 8$$

then

$$8 = 2 \cdot 2 \cdot 2 = 2^3$$

The base 2 is used as a factor three times, therefore the exponent is 3.

EXAMPLE 4
continued

$$16 = 2 \cdot 2 \cdot 2 \cdot 2 = 2^4$$

EXAMPLE 5

The base 2 is used as a factor four times, therefore the exponent is 4.

The decimal system of numeration is a "base ten" system, therefore, 10 and the powers of 10 are important.

$$10 = 10^1$$
$$100 = 10 \cdot 10 = 10^2$$
$$1\ 000 = 10 \cdot 10 \cdot 10 = 10^3$$
$$10\ 000 = 10 \cdot 10 \cdot 10 \cdot 10 = 10^4$$
$$\frac{1}{100} = \frac{1}{10 \cdot 10} = \frac{1}{10^2} \text{ (which also may be written } 10^{-2})$$
$$\frac{1}{1000} = \frac{1}{10 \cdot 10 \cdot 10} = \frac{1}{10^3} \text{ (which also may be written } 10^{-3})$$

10^1, 10^2, 10^3, 10^4, $\frac{1}{10}$ or 10^{-1}, $\frac{1}{10^2}$ or 10^{-2}, $\frac{1}{10^3}$ or 10^{-3}, and so on are called *powers of 10*.

> A close look at the powers of 10 discloses that the exponent on the 10 indicates the number of zeros in the product form of the answer. This special relationship between the power and the base is true only when the base is 10.

$$10^2 = 100$$

The exponent is 2 and there are two zeros in 100.

$$10^3 = 1\ 000$$
$$10^4 = 10\ 000$$
$$10^5 = 100\ 000$$
$$10^6 = 1\ 000\ 000$$
$$\frac{1}{10^4} = \frac{1}{10\ 000}$$

Exercises 2.1   Translate from exponential form to factored form ($10^2 = 10 \cdot 10$).

     1. 10^1 2. 10^3

     3. 10^4 4. 10^5

     5. $\dfrac{1}{10^3}$ 6. $\dfrac{1}{10^4}$

Write in product form ($10^2 = 100$).

     7. 10^1 8. 10^3

     9. 10^4 10. 10^5

    11. $\dfrac{1}{10^3}$ 12. $\dfrac{1}{10^4}$

Write in exponential form as a power of 10 ($100 = 10^2$).

    13.   10 000   14.   1000

    15.   10   16. $\dfrac{1}{100}$

    17. $\dfrac{1}{1000}$ 18. $\dfrac{1}{100\ 000}$

Section 2.2   **Reading and Writing Decimal Numerals**

The word *decimal* comes from the Latin word *decimus,* which means *tenth.* The decimal system of numeration is a "base ten" system. To better understand the system, study these numerals written in expanded form:

EXAMPLE 1

$$368 = 300 + 60 + 8$$
$$= 3(100) + 6(10) + 8(1)$$
$$= 3(10^2) + 6(10^1) + 8(1)$$

EXAMPLE 2

$$4572 = 4000 + 500 + 70 + 2$$
$$= 4(1000) + 5(100) + 7(10) + 2(1)$$
$$= 4(10^3) + 5(10^2) + 7(10^1) + 2(1)$$

$$625.43 = 600 + 20 + 5 + \frac{4}{10} + \frac{3}{100}$$

EXAMPLE 3

$$= 6(100) + 2(10) + 5(1) + 4\left(\frac{1}{10}\right) + 3\left(\frac{1}{100}\right)$$

$$= 6(10^2) + 2(10^1) + 5(1) + 4\left(\frac{1}{10^1}\right) + 3\left(\frac{1}{10^2}\right)$$

Note that when a decimal numeral is written in expanded form, each position has a power of 10 factor. This is the basis for naming the decimal system of numeration a "base ten" system of numeration. This fact and the names for each of the powers of 10 provide a method for reading and writing decimal numerals. Use Table 2.1 for position names as you study these numerals and their names:

$$0.6 = \text{six-tenths}$$
$$0.4009 = \text{four thousand nine ten-thousandths}$$
$$8000.007 = \text{eight thousand and seven-thousandths}$$

Whole number part							decimal point	Decimal fraction part					
million	hundred thousand	ten thousand	thousand	hundred	ten	one	·	tenth	hundredth	thousandth	ten thousandth	hundred thousandth	millionth
10^6	10^5	10^4	10^3	10^2	10^1	1	·	$\frac{1}{10}$	$\frac{1}{10^2}$	$\frac{1}{10^3}$	$\frac{1}{10^4}$	$\frac{1}{10^5}$	$\frac{1}{10^6}$

A second method for reading and writing decimal numerals stems from the fact that decimal fractions when expressed in common fraction form have as denominators 10 or a power of 10.

| **EXAMPLE 4** | A decimal fraction with one position to the right of the decimal point indicates one zero in the power of 10 denominator $\left(\frac{1}{10}\right)$ and is read *tenth(s)*: |

$$0.5 = \frac{5}{10} = \text{five-tenths}$$

| **EXAMPLE 5** | A decimal fraction with two positions to the right of the decimal point indicates two zeros in the power of 10 denominator $\left(\frac{1}{100}\right)$ and is read *hundredth(s)*: |

$$0.65 = \frac{65}{100} = \text{sixty-five hundredths}$$

| **EXAMPLE 6** | A decimal with three positions to the right of the decimal point indicates three zeros in the power of 10 denominator $\left(\frac{1}{1000}\right)$ and is read *thousandth(s)*: |

$$0.005 = \frac{5}{1000} = \text{five-thousandths}$$

> The number of positions to the right of the decimal point in a decimal fraction indicates the number of zeros in the power of 10 denominator in the equivalent common fraction. This power of 10 denominator names the decimal numeral.

Some Observations About the Decimal System

1. When there is no whole number part in the decimal numeral—that is, the numeral has a value less than 1—write a zero in the ones' position:

   0.006   0.297   0.5   0.6   0.68

2. To read a decimal numeral that has a whole number part and a fraction part, indicate the decimal point verbally by the word *and*. Be sure to use *and* only when you want to indicate a decimal point.

3.5 = *three* and *five-tenths*

2004.06 = *two thousand four* and *six-hundredths*

3. When there is no written decimal point in a numeral, the decimal point is understood to be at the right of the numeral.

$$2 = 2. \qquad 352 = 352. \qquad 4795 = 4795.$$

4. Zero attached to a numeral does not alter the value of the numeral so long as the attached zero is not placed between the decimal point and another number.

$$3.5 \ = 3.50, \ \text{but } 3.5 \ \neq 3.05$$
$$3.52 = 3.520, \text{but } 3.52 \neq 3.502$$

5. Numerals containing more than four digits are grouped by three digits going left and right from the decimal point. By international agreement, a comma is no longer used.

1 200 000 not 1,200,000

67 432.123 45 not 67,432.12345

6. Each position has a place value that is ten times greater than the value of the position immediately on its right.

$$268.75 = 2(100) + 6(10) + 8(1) + 7\left(\frac{1}{10}\right) + 5\left(\frac{1}{100}\right)$$

100 is 10 times 10, which is 10 times 1, which is 10 times $\frac{1}{10}$, which is 10 times $\frac{1}{100}$.

Write in expanded notation form. For example, $24.5 = 2(10) +$ **Exercises 2.2**
$4(1) + 5(\frac{1}{10})$.

1. 482   2. 245
3. 3005   4. 90 001
5. 6.74   6. 4.005

Write as decimal numerals.

7. six and seven-tenths   8. six and seven-thousandths
9. sixty-seven millionths   10. six hundred seven and sixty-seven ten-thousandths

11. $\dfrac{6}{100}$ 12. $2\dfrac{15}{1000}$

13. $\dfrac{165}{100\,000}$ 14. $\dfrac{707}{10\,000}$

15. $23\dfrac{48}{1\,000\,000}$

Translate the fractions into words.

16. 0.003   17. 8.000 003

18. 4.305   19. 6.74

20. 0.70   21. 0.100 78

22. 700 000.000 008   23. $\dfrac{165}{100\,000}$

24. $23\dfrac{48}{1\,000\,000}$ 25. 0.01

26. 0.001   27. 0.0001

Section 2.3   Addition and Subtraction

Addition and subtraction of decimal fractions are much like the addition and subtraction of whole numbers.

1. Both operations are easier if numerals are written in column form.

$$3.6 + 0.2 = \quad \begin{array}{r} 3.6 \\ +\,0.2 \\ \hline 3.8 \end{array}$$

$$3.6 - 0.2 = \quad \begin{array}{r} 3.6 \\ -\,0.2 \\ \hline 3.4 \end{array}$$

2. It is very important that the decimal points align, forming a single vertical column.

$$50 + 0.8 + 3.059 = \quad \begin{array}{r} 50.000 \\ 0.800 \\ +\;\;3.059 \\ \hline 53.859 \end{array}$$

$$2.78 - 1.5 = \quad \begin{array}{r} 2.78 \\ -\ 1.50 \\ \hline 1.28 \end{array}$$

3. The decimal point in your answer is placed in the same vertical decimal point column.

4. Zeros can be placed to the right of the numerals so that each numeral has the same number of digits to the right of the decimal point. This is optional with addition but may be required for subtraction.

$$50.07 + 0.8 + 3.059 = \quad \begin{array}{r} 50.07\underline{0} \text{ (optional zero)} \\ 0.8\underline{00} \text{ (optional zeros)} \\ +\ \ 3.059 \\ \hline 53.929 \end{array}$$

$$2.78 - 1.5 = \quad \begin{array}{r} 2.78 \\ -\ 1.5\underline{0} \text{ (optional zero)} \\ \hline 1.28 \end{array}$$

$$0.35 - 0.062 = \quad \begin{array}{r} 0.35\underline{0} \text{ (zero is necessary)} \\ -\ 0.062 \\ \hline 0.288 \end{array}$$

5. The operations of addition and subtraction are performed as with whole numbers, simply keeping the decimal points aligned.

In Exercises 1–20 add or subtract as indicated.   **Exercises 2.3**

1. $6.92 + 10.07 + 1.1 =$
2. $2.003 + 12.9 + 67 =$
3. $16.35 - 8.2 =$
4. $16.2 - 8.356 =$
5. $9.002 - 4.3 =$
6. $19.003 - 12.5 =$
7. $100.17 - 0.117 =$
8. $200.48 - 0.448 =$
9. $10.08 + 3.047 =$
10. $0.29 + 0.029 + 3.46 + 15 =$

11. $35.009 + 0.009 + 5.4 =$
12. $20.05 + 8.208 =$
13. $17.0015 - 0.406 =$
14. $10.0008 - 0.707 =$
15. $10.1010 - 1.0101 =$
16. $6 - 0.067 =$
17. $0.892 + 6.75 + 3 =$
18. $6.006 + 0.09 + 1.1 =$
19. $15 + 0.008 + 9.43 =$
20. $17.0015 + 0.6 + 5.09 =$

21. 0.2 gram, 0.5 gram, 1 gram, and 5 grams are a total of how many grams?

Section 2.4   Multiplication

Three common forms are used to indicate multiplication of decimal numerals.

$$(0.25)\,(0.7) \qquad 0.25 \times 0.7 \qquad 0.25 \cdot 0.7$$

The first and second forms are preferred. Multiplication of decimal numerals is much like multiplication of whole numbers once the placement of the decimal point in the product is understood. We can use common fractions to illustrate.

EXAMPLE 1

Multiply 0.25 by 0.7. To begin, translate the decimal numerals to common fractions (Section 2.1). Remember, two positions to the right of the decimal point in 0.25 means two zeros in the denominator; one position to the right of the decimal point in 0.7 means one zero in the denominator.

$$0.25 \times 0.7 = \frac{25}{100} \times \frac{7}{10}$$

Do the multiplication:

$$= \frac{175}{1000}$$

Now translate the product to a decimal numeral.

$$= 0.175$$

There are two zeros in the denominator of $\left(\frac{25}{100}\right)$ plus one zero in the denominator $\left(\frac{7}{10}\right)$ for a total of three zeros, which when written as a decimal numeral will have three positions to the right of the decimal point.

To place the decimal point in the product, remember that the number of positions to the right of the decimal point in the answer is equal to the sum of the number of positions to the right of the decimal point in each factor. Once this is understood, we can multiply as with whole numbers and place the decimal point as follows.

EXAMPLE 2

$$0.25 \times 0.7 =$$

0.25	two positions right of the decimal point
× 0.7 +	one position right of the decimal point
0.175	three positions right of the decimal point

Add the number of positions used to the right of the decimal point and use their total to position the decimal point in the product.

EXAMPLE 3

$$(2.06)(0.854) =$$

2.06	two positions right of the decimal point
×0.854 +	three positions right of the decimal point
824	
1030	
1648	
1.75924	five positions right of the decimal point

EXAMPLE 4

$$0.005 \times 0.08 =$$

0.005	three positions right of the decimal point
×0.08 +	two positions right of the decimal point
0.00040	five positions right of the decimal point

Note that the zero to the right of the 4 counts as a position because 5 times 8 equals 40.

Multiplication by 10 and the Powers of 10

When you are multiplying by 10 and powers of 10 that are greater than 1, the decimal point shifts to the right as many positions as there are zeros in the 10 or power of 10.

| EXAMPLE 5 | $2.5 \times 10 = 25$ |

There is one zero in 10; therefore, the decimal point shifts one place to the right.

| EXAMPLE 6 | $2.5 \times 100 = 250$ |

There are two zeros in 100; therefore, the decimal point shifts two places to the right. If the decimal fraction does not have enough positions, add zeros on the right as you need them. Also see examples 7 through 9.

| EXAMPLE 7 | $0.76 \times 1000 = 760$ |

There are three zeros in 1000; therefore, the decimal point shifts three places to the right.

| EXAMPLE 8 | $0.005 \times 10\ 000 = 50$ |

There are four zeros in 10 000; therefore, the decimal point shifts four places to the right.

| EXAMPLE 9 | $32 \times 10 = 320$ |

Examples 5 through 9 all show multiplication involving 10 or the power of 10 in product form. The same theory applies even when the power of 10 is written in exponential form.

| EXAMPLE 10 | $0.06 \times 10^2 = 6$ |

There are two zeros in 10^2; therefore, the decimal point shifts two places to the right.

| EXAMPLE 11 | $6.1 \times 10^3 = 6100$ |

There are three zeros in 10^3; therefore, the decimal point shifts three places to the right.

$$0.0549 \times 10^4 = 549$$

There are four zeros in 10^4; therefore, the decimal point shifts four places to the right.

EXAMPLE 12

> One formula that frequently is used to determine dosage involves multiplication of decimal numerals.
>
> Body surface area in square metres $\times$ dose per square metre
>
> = Desired medication

A child with a body surface area of 0.6 square metre is to be given epinephrine 0.3 millilitre per square metre body surface. How much epinephrine should be administered per dose?

EXAMPLE 13

$$0.6 \times 0.3 \text{ millilitre} = 0.18 \text{ millilitre epinephrine per dose}$$

Find the following products.

1. $0.68 \times 0.345 =$
2. $16.08 \times 3.57 =$
3. $0.0085 \times 100 =$
4. $0.543 \times 0.07 =$
5. $1.68 \times 0.1 =$
6. $3.79 \times 2.1 =$
7. $0.6 \times 0.006 =$
8. $15.12 \times 1000 =$
9. $135.6 \times 2.04 =$
10. $0.0084 \times 10 =$
11. $4.58 \times 10^4 =$
12. $6.003 \times 10^2 =$
13. $0.015 \times 0.009 =$
14. $0.54 \times 0.01 =$
15. $(7.39)(1.5) =$
16. $(0.008)(0.14) =$
17. $(7.008)(10^3) =$
18. $(3.01)(242.7) =$
19. $(6.854)(0.001) =$
20. $(3.428)(15.69) =$
21. The doctor asks you to administer 1.5 milligrams of a drug three times daily. What is the total amount administered daily? Weekly?
22. 0.6 of .035 grams is how many grams?
23. 0.02 of 6.4 millilitres is how many millilitres?

24.   A child who has a body surface area of 0.72 square metre is to be given sodium phosphate NF 2.5 grams per square metre body surface. How much sodium phosphate NF will you administer per dose?

25.   A child who has a body surface area of 1.06 square metres is to be given diazepam NF 3.5 milligrams per square metre body surface. How much diazepam will you administer per dose? If the dose is repeated four times daily, what will be the total amount administered daily?

Section 2.5   Division

Division involving decimal numerals is much like the division of whole numbers, but again special attention must be paid to the placement of the decimal point. First, apply the Identity Property of Multiplication (Section 1.3) to the numerals in order to obtain a whole number divisor.

EXAMPLE 1

$$8.4 \div 0.2 = \frac{8.4}{0.2}$$

0.2 is the divisor. To convert 0.2 to a whole number, multiply it by the name for $1\left(\frac{10}{10}\right)$ that will result in the whole number 2.

$$= \frac{8.4}{0.2} \cdot \frac{10}{10}$$

$$= \frac{84}{2}$$

EXAMPLE 2

$$8.4 \div 0.002 = \frac{8.4}{0.002}$$

Choose $\frac{1000}{1000}$ as the name for 1 because it gives the whole number 2 as a divisor.

$$= \frac{8.4}{0.002} \cdot \frac{1000}{1000}$$

$$= \frac{8400}{2}$$

$$0.084 \div 0.02 = \frac{0.084}{0.02}$$

EXAMPLE 3

Choose $\frac{100}{100}$ as the name for 1 because it gives the whole number 2 as a divisor.

$$= \frac{0.084}{0.02} \cdot \frac{100}{100}$$

$$= \frac{8.4}{2}$$

Once the divisor is converted to a whole number, proceed with the division as you would whole number division, taking care to observe two things:

1. Place the decimal point in the quotient directly above the decimal point in the dividend.
2. Keep the columns straight.

$$8.4 \div 0.2 = \frac{8.4}{0.2}$$

EXAMPLE 4

$$= \frac{8.4}{0.2} \cdot \frac{10}{10}$$

$$= \frac{84}{2}$$

$$= 2\overline{)84.}\overset{\displaystyle 42.}{}$$

$$\underline{84}$$

$$0.084 \div 0.02 = \frac{0.084}{0.02}$$

EXAMPLE 5

$$= \frac{0.084}{0.02} \cdot \frac{100}{100}$$

$$= \frac{8.4}{2}$$

EXAMPLE 5
continued

$$= 2\overline{)8.4} \atop 8\ 4$$ with quotient 4.2

EXAMPLE 6

$$3 \div 6.4 = \frac{3.}{6.4}$$

$$= \frac{3}{6.4} \cdot \frac{10}{10}$$

$$= \frac{30}{64}$$

$$= 64\overline{)30.00000}$$ with quotient 0.46875

$$
\begin{array}{r}
25\ 6 \\ \hline
4\ 40 \\
3\ 84 \\ \hline
560 \\
512 \\ \hline
480 \\
448 \\ \hline
320 \\
320 \\ \hline
\end{array}
$$

Note that zeros may be added to the dividend as needed.

Remember that the verbal statements *is what part of* and *is how many times* each indicates the operation of division (Section 1.7).

Division by 10 and the Powers of 10

To divide by 10 or a power of 10, shift the decimal point in the dividend to the left as many positions as there are zeros in the 10 or power of 10 divisor.

EXAMPLE 7

$$25 \div 10 = 2.5 \qquad \frac{0.68}{10} = 0.068$$

There is one zero in 10; therefore, the decimal point shifts one place to the left.

EXAMPLE 8

$$25 \div 100 = 0.25 \qquad \frac{7.008}{100} = 0.070\ 08$$

There are two zeros in 100; therefore, the decimal point shifts two places to the left.

EXAMPLE 9

$$25 \div 1000 = 0.025 \qquad \frac{7.008}{1000} = 0.007\ 008$$

There are three zeros in 1000; therefore, the decimal point shifts three places to the left.

Rounding Off

Example 6 in this section brings up the question of quotients and their accuracy. *In calculations, one need be only as accurate as the equipment being used.* For example, to administer a dose of less than 1 millilitre a tuberculin syringe may be used. A tuberculin syringe is accurate to the hundredth place (two decimal positions to the right of the decimal point). *Therefore, calculations for a dose of less than 1 millilitre that is to be administered in a tuberculin syringe should be accurate only to the hundredth place.* Handling this type of calculation requires an understanding of rounding off solutions to dosage problems.*

In order to round off decimal fractions, place your pencil on the position you have been asked to round to. Look to the right of that position.

---

*Instructions for rounding off generally will appear in one of two forms: *round to the hundredth place* or *accurate to the hundredth place.*

1. For numbers less than 1, if the digit on the right is less than 5, drop the digit (and any that follow it) and leave the rest of the numeral unaltered.

EXAMPLE 10

Round 2.364 to the hundredth place.

$$2.364 = 2.36$$
$$\uparrow$$

6 is in the hundredth place. 4 on the right is less than 5, so drop the 4 and leave 2.36 unchanged.

EXAMPLE 11

Round 7.5423 to the thousandth place.

$$7.5423 = 7.542$$
$$\uparrow$$

2 is in the thousandth position. 3 on the right is less than 5, so drop the 3 and leave 7.542 unchanged.

2. For the numbers less than 1, if the digit on the right is greater than 5, drop the digit and add 1 to the round-to position.

EXAMPLE 12

Round 2.397 to the hundredth place.

$$2.397 = 2.40$$
$$\uparrow$$

9 is in the hundredth position. 7 on the right is greater than 5, so drop the 7 and add 1 to 9. *The zero is retained to show that the answer is correct to the hundredth place.*

EXAMPLE 13

Round 7.98 to the tenth place.

$$7.98 = 8.0$$
$$\uparrow$$

9 is in the tenth place. 8 on the right is greater than 5, so drop the 8 and add 1 to 9. The zero in the tenth position is retained to show that the answer is correct to that position.

EXAMPLE 13
continued

3. For numbers less than 1, if the digit on the right is exactly 5 (or 50, or 500, or so on) with no remainder, drop the 5 (or 50, or 500, or so on) and round to the even number.

Round 2.365 to the hundredth place.

$$2.365 = 2.36$$
$$\uparrow$$

EXAMPLE 14

6 is in the hundredth place. 5, the digit to its right, is exactly halfway between 2.36 and 2.37. Therefore, choose the even number 2.36.

Round 2.375 to the hundredth place.

$$2.375$$
$$\uparrow$$

EXAMPLE 15

7 is in the hundredth place. 5 on the right is exactly halfway between 2.37 and 2.38. Therefore, choose the even number 2.38.

Obviously in order to give an answer correct to a certain position, you must continue the division to at least one position to the right of the requested position and then round to the requested position. That is,

- to give an answer correct to the tenth place, divide to the hundredth place and then round to the tenth place.

- to give an answer correct to the hundredth place, divide to the thousandth place and then round to the hundredth place.

Some Symbols Used in Division

There are three types of decimal fractions encountered in quotients.

1. The *terminating decimal,* in which the decimal numeral has a finite number of digits to the right of the decimal point.
2. The *repeating decimal,* in which the decimal numeral has an unending number of digits that repeat in a fixed pattern to the right of the decimal point.
3. The *nonrepeating and nonterminating decimal,* in which the decimal numeral has an unending number of digits to the right of the decimal point but these digits never repeat and never terminate.

Rounding off may be used for any of these types of decimal fractions. Even a terminating decimal may terminate at a place beyond the round-to position. There are two symbols used to indicate the second type of fraction, the repeating decimal.

1. Three dots following a decimal fraction indicate to continue in the same pattern; that is, the pattern repeats infinitely.

    8.333 . . . (the 3 repeats infinitely)

    8.1212 . . . (the 12 repeats infinitely)

2. A bar over the repeating part of a decimal fraction indicates that that part under the bar continues to repeat infinitely.

    $8.\overline{3}$ (the 3 repeats infinitely)

    $8.\overline{12}$ (the 12 repeats infinitely)

Exercises 2.5

Give the following correct to the tenth position:

1. 0.333 . . .   2. $0.\overline{6}$
3. 2.04   4. 0.586
5. 5.55   6. 3.09

Give the following correct to the hundredth position:

7. 0.333 . . .   8. $0.\overline{6}$
9. 7.387   10. 0.3738
11. 9.999   12. 6.885

Give the following correct to the thousandth position:

13. 0.333 . . .   14. $0.\overline{6}$

15. 4.5866   16. 0.543 49

17. 13.0005   18. 6.4998

Do the division in Exercises 19–38. Your answers should be correct to the thousandth place.

19. $0.64 \div 3 =$ 20. $0.05 \div 45 =$

21. $3 \div 0.64 =$ 22. $62 \div 0.004 =$

23. $0.64 \div 0.2 =$ 24. $0.006 \div 10 =$

25. $0.2 \div 0.64 =$ 26. $3.6 \div 9 =$

27. $307.52 \div 10\ 000 =$ 28. $3.6 \div 0.009 =$

29. $0.562 \div 0.32 =$ 30. $3.6 \div 0.9 =$

31. $89.9 \div 0.43 =$ 32. $7 \div 92 =$

33. $45.3 \div 10^2 =$ 34. $50.085 \div 90.01 =$

35. $135 \div 2.2 =$ 36. $158 \div 2.2 =$

37. $0.7 \div 10^3 =$ 38. $178.4 \div 0.033 =$

You may find a review in Section 1.7 will help with the following problems.

39. 2.05 grams is what part of 6.15 grams?

40. 0.006 litre is what part of 0.18 litre?

41. 0.12 millilitre is how many times 0.2 millilitre?

42. 0.18 litre is how many times 0.006 litre?

Comparing Sizes of Decimal Numerals   Section 2.6

To compare the sizes of two or more decimal numerals, follow this simple procedure:

1. Write the numerals in column form, keeping the decimal points in a single vertical column.

2. Fill in with zeros until each numeral has the same number of digits to the right of the decimal point.

3. Ignore the decimal point and read the numerals as though they were whole numbers.

EXAMPLE 1	Which is smaller, 0.1 or 0.06?

$$0.1$$
$$0.06 \tag{1}$$

Align the numerals in a column, and fill in with zeros:

$$0.1\underline{0}$$
$$0.06 \tag{2}$$

Reading the numerals as whole numbers, 6 is smaller than 10. Therefore,

$$0.06 < 0.1 \tag{3}$$

EXAMPLE 2	Arrange 0.0099, 0.06, 0.1, 9.025, 0.0006, 0.648, and 7.64 from largest to smallest. Begin by writing the numbers in column form, keeping the decimals in a single vertical column:

0.0099

0.06

0.1

9.025

0.0006

0.648

7.64

Now fill in the blank positions to the right of the decimal point in each numeral until each column is complete, or until each numeral has the same number of digits to the right of the decimal point.

0.0099

0.06$\underline{00}$

0.1$\underline{000}$

9.025$\underline{0}$

0.0006

0.648$\underline{0}$

7.64$\underline{00}$

Now, reading them as whole numbers (99, 600, 1000, 90 250, 6, 6480, and 76 400) arrange them in descending order:

$$9.025 > 7.64 > 0.648 > 0.1 > 0.06 > 0.0099 > 0.0006$$

Use this same procedure to check equality.

EXAMPLE 3

Is 0.5 the same as 0.50?

0.5

0.50

0.5<u>0</u>

0.50

50 equals 50, so

$$0.5 = 0.50$$

In Exercises 1–10, make each statement true by placing =, >, or <   **Exercises 2.6**
between each pair of numbers.

1. 0.02   0.0025   2. 0.105   0.0994

3. 0.885   1.00   4. 0.000 95   0.0043

5. 0.040   0.04   6. 3.007   3.007 00

7. 0.105   0.0105   8. 10.003   10.030

9. $\dfrac{1}{1000}$ 0.100   10. 0.6 $\dfrac{6}{100}$

Arrange the numbers in Exercises 11 and 12 in order from smallest
to largest.

11. 0.3   0.047   0.03   0.33   0.003   0.157

12. 6.008   0.105   0.0099   0.6008   0.06   0.10001

13.   The doctor ordered 0.0035 gram of a drug. You gave 0.004
      gram. Did you give the correct amount? If not, did you give
      too much or too little?

Decimal Numerals ⇌ Common Fractions

Section 2.7

An easy way to change a decimal numeral to a common fraction is
to read the decimal numeral. Its name tells you the common fraction (see Section 2.2).

**Decimal
Numerals to
Common
Fractions**

EXAMPLE 1

$$0.657 = \text{six hundred fifty-seven } \textit{thousandths}$$
$$= \frac{657}{1000}$$

EXAMPLE 2

$$0.0005 = \text{five } \textit{ten-thousandths}$$
$$= \frac{5}{10\ 000}$$

EXAMPLE 3

$$3.02 = \text{three and two-}\textit{hundredths}$$
$$= 3\frac{2}{100}$$

Once you've converted to the power of 10 common fraction, you can reduce the common fraction if you like (Section 1.3).

$$0.0005 = \frac{5}{10\ 000} = \frac{1}{2000}$$
$$3.02 \quad = 3\frac{2}{100} = 3\frac{1}{50}$$

> Another useful idea to recall is that when you change a decimal numeral to a common fraction there will be as many zeros in the power of 10 denominator in the common fraction as there are positions to the right of the decimal point in the decimal numeral (Section 2.2).

EXAMPLE 4

$$0.45 = \frac{45}{100}$$

There are two positions to the right of the decimal point in 0.45, so there are two zeros in the power of 10 denominator $\left(\frac{}{100}\right)$.

$$6.00207 = 6\frac{207}{100\ 000}$$

EXAMPLE 5

There are five positions to the right of the decimal point in 6.002 07, so there are five zeros in the power of 10 denominator $\left(\frac{}{100\ 000}\right)$.

Two different methods are used to change common fractions to decimal numerals. First, in those cases where the denominator of the common fraction easily can be changed to a power of 10 denominator, apply the Identity Property of Multiplication (Section 1.3) and do so. Then convert the common fraction to a decimal numeral by applying the rule for division by 10, or powers of 10 (Section 2.5).

$$\frac{1}{2} = \frac{1}{2} \cdot \frac{5}{5} = \frac{5}{10} = 0.5$$

$$\frac{3}{4} = \frac{3}{4} \cdot \frac{25}{25} = \frac{75}{100} = 0.75$$

$$\frac{1}{25} = \frac{1}{25} \cdot \frac{4}{4} = \frac{4}{100} = 0.04$$

$$\frac{3}{125} = \frac{3}{125} \cdot \frac{8}{8} = \frac{24}{1000} = 0.024$$

EXAMPLE 6

In those instances where it is not easy to change the denominator of the common fraction to a power of 10, simply divide the denominator into the numerator. Your quotient is the equivalent decimal fraction.

Change $\frac{1}{3}$ to a decimal numeral.

EXAMPLE 7

$$\frac{1}{3} = 1 \div 3$$

$$= 3\overline{)1.000}^{\ 0.333\ ...}$$

Note that the quotient may appear as

> 0.333 . . . (the three dots indicate to continue in the same pattern), or

$0.\overline{3}$ (the bar over the 3 indicates a repeating 3), or

~~0.3 (rounded to the tenth place), or~~

0.33 (rounded to the hundredth place), or so on.

EXAMPLE 8   Change $\frac{1}{6}$ to a decimal numeral.

$$\frac{1}{6} = 1 \div 6 = 6\overline{\smash{)}1.0000}$$

$$\begin{array}{r}
0.1666\ldots \\
\underline{6} \\
40 \\
\underline{36} \\
40 \\
\underline{36}
\end{array}$$

Again the quotient may be left as

0.1666 . . . (the three dots indicate that the last three numerals repeat in the same pattern) or

$1.1\overline{6}$ (the bar indicates that the 6 repeats), or you could round off to

0.2 (the tenth place), to

0.17 (the hundredth place), to

0.167 (the thousandth place), or so on.

Both methods also work with improper fractions.

EXAMPLE 9   Change $\frac{8}{5}$ to a decimal numeral.

$$\frac{8}{5} = \frac{8}{5} \cdot \frac{20}{20}$$

$$= \frac{160}{100}$$

$$= 1.6$$

EXAMPLE 10   Change $\frac{87}{4}$ to a decimal numeral.

$$\frac{87}{4} = 87 \div 4$$

$$= 4\overline{)\begin{array}{r} 21.75 \\ 87.00 \end{array}}$$

EXAMPLE 10
continued

$$\begin{array}{r} 84 \\ \hline 30 \\ 28 \\ \hline 20 \\ 20 \\ \hline \end{array}$$

Any common fraction when converted to a decimal numeral will either divide evenly into its numerator as in Examples 9 and 10 (a terminating decimal) or will make a repeating decimal, as in Examples 7 and 8.

Change the decimal numerals in Exercises 1–10 to equivalent common fractions, reducing if possible.   **Exercises 2.7**

1. 0.6   2. 0.06
3. 0.0006   4. 9.025
5. 0.675   6. 0.000 025
7. 6.5   8. 0.1025
9. 0.10   10. 0.100 04

Change the following common fractions to equivalent decimal numerals.

11. $\frac{3}{4}$ 12. $\frac{11}{20}$

13. $\frac{1}{7}$ 14. $\frac{5}{12}$

15. $\frac{8}{5}$ 16. $\frac{21}{4}$

17. $\frac{21}{125}$ 18. $\frac{5}{6}$

19. $\frac{2}{9}$ 20. $\frac{3}{8}$

Place $=$, $>$, or $<$ between each pair of numbers to make the statements true.

21. $\dfrac{1}{2}$ 0.50   22. 0.67 $\dfrac{2}{3}$

23. 0.3 $\dfrac{1}{3}$ 24. $\dfrac{2}{7}$ 0.2857

25. $\dfrac{7}{6}$ 1.2   26. $\dfrac{1}{8}$ 0.0125

Review Exercises   **[2.1]**

1. Convert to factored form.

   a. 10^5 b. $\dfrac{1}{10^3}$

2. Convert to exponential form.

   a.   100 000   b. $\dfrac{1}{10\ 000}$

3. Convert to product form.

   a. 10^3 b. $\dfrac{1}{10^2}$

[2.2]

4. Write in expanded form.
   a.   348   b.   7.0408

5. Write as a decimal numeral.
   a.   one hundred eight and thirteen-thousandths

   b. $2\dfrac{60}{10\ 000}$

6. Write in words.
   a.   0.009   b.   200.0109

[2.3]

7. Add or subtract as indicated.
   a. $10.003 + 0.01 + 21 =$ b. $0.2 + 3 + 0.078$
                                          $+ 14.329 =$

   c. $17 - 8.509 =$ d. $9.006 - 0.25 =$

[2.4]

8. Find the following products.

   a. $3.06 \times 0.05 =$ b. $0.72 \times 0.002 =$

   c. $6.05 \times 10^4 =$ d. $0.0089 \times 1000 =$

9. 0.6 of 3.5 millilitres is how many millilitres?

10. 0.02 of 25.5 kilograms is how many kilograms?

[2.5]

11. Round 3.9045 to the indicated position:

   a.   thousandth place   b.   hundredth place

   c.   tenth place   d.   ones' place

12. Find the following quotients correct to the thousandth place.

   a. $0.063 \div 0.3 =$ b. $0.3 \div 0.063 =$

   c. $5.94 \div 10^2 =$ d. $75 \div 10\,000 =$

13. 0.063 is what part of 0.3?

14. 0.3 is how many times 0.063?

[2.6]

15. Place $=$, $>$, or $<$ between each pair of numbers to make a true statement.

   a.   0.04   0.0045   b.   6.06   6.060

[2.7]

16. Change to equivalent common fractions.

   a.   0.104   b.   70.08

17. Change to equivalent decimal numerals.

   a. $\dfrac{5}{11}$ b. $\dfrac{17}{25}$

18. Place $=$, $>$, or $<$ between each pair of numbers to make a true statement.

   a.   0.03 $\dfrac{1}{30}$ b. $\dfrac{25}{4}$ 6.24

   c.   0.125 $\dfrac{1}{8}$ d.   0.33 $\dfrac{1}{3}$

CHAPTER

3

Percentages

Percentage is a convenient method for representing *the part of a whole expressed in hundredths*. Percentages are widely used in many disciplines including the health professions. Two health profession examples come to mind: it is common for drug concentration to be expressed as a percentage (0.5% solution); percentage may be used in the calculation of desired medication (administer a drug in the amount of 2% of body mass or weight). Because of its extensive use, the topic of percentages is discussed in detail in this chapter.

Topics include:

Definition of percentage or percent

Conversion of percentages to common fractions, and decimal numerals, and vice versa

Operations of addition, subtraction, multiplication, and division involving mixtures of common fractions, decimal numerals, and percentages

Verbal expressions indicating multiplication, division, or percentage

1. Change 5.5% to

    a.   an equivalent decimal numeral.

    b.   an equivalent common fraction in lowest terms.

2. Change 0.02% to

    a.   an equivalent decimal numeral.

    b.   an equivalent common fraction in lowest terms.

3. Change 150% to

    a.   an equivalent decimal numeral.

    b.   an equivalent common fraction in lowest terms.

[3.2]

4. Change the following common fractions and decimal numerals to percentages.

    a. $\dfrac{1}{40}$ b. $\dfrac{10}{3}$

    c.   0.0067   d.   0.105

    e.   2.5   f. $1\dfrac{5}{8}$

[3.3]

5. Perform the indicated operations.

    a. $\dfrac{1}{4} + 1.03 =$ b. $0.75 - \dfrac{1}{6} =$

    c. $50\% + 2 =$ d. $1.95 \times \dfrac{1}{3} =$

    e. $\dfrac{3}{2} \div 0.75 =$ f. $12\dfrac{1}{2}\%$ of $64 =$

    g.   0.3% of 0.3 =   h. $15 \div 1.5\% =$

Converting to Common Fractions and Decimal Numerals

Section 3.1

The percent form of writing numerals has been in widespread use for many hundreds of years. A percentage is a division by 100.

To Common Fractions

Fifty percent means *50 divided by 100*, or $\frac{50}{100}$. In fact, the percent symbol (%) is defined as *division by 100*.

$$3\% \text{ means } 3 \text{ divided by } 100, \text{ or } \frac{3}{100}$$

$$47\% \text{ means } 47 \text{ divided by } 100, \text{ or } \frac{47}{100}$$

$$99\% \text{ means } 99 \text{ divided by } 100, \text{ or } \frac{99}{100}$$

Obviously, changing a percentage to a common fraction is a simple procedure. Once in common fraction form, it is possible to reduce the common fraction to its simplest form (Section 1.3).

$$2\% = \frac{2}{100} = \frac{1}{50}$$

$$15\% = \frac{15}{100} = \frac{3}{20}$$

$$150\% = \frac{150}{100} = \frac{3}{2} = 1\frac{1}{2}$$

$$200\% = \frac{200}{100} = \frac{2}{1} = 2$$

In some instances the numerator will not be a whole number. To convert to a whole number numerator, use one of the following methods:

1. If the numerator is a decimal numeral, apply the Identity Property of Multiplication, using the appropriate name for 1, to make a whole number numerator.

$$0.02\% = \frac{0.02}{100} = \frac{0.02}{100} \times \frac{100}{100} = \frac{2}{10\,000} = \frac{1}{5000}$$

$$1.5\% = \frac{1.5}{100} = \frac{1.5}{100} \times \frac{10}{10} = \frac{15}{1000} = \frac{3}{200}$$

2. If the numerator is a mixed number, perform the division to simplify.

$$12\frac{1}{2}\% = \frac{12\frac{1}{2}}{100} = 12\frac{1}{2} \div 100 = \frac{25}{2} \cdot \frac{1}{100} = \frac{1}{8}$$

$$33\frac{1}{3}\% = \frac{33\frac{1}{3}}{100} = 33\frac{1}{3} \div 100 = \frac{100}{3} \cdot \frac{1}{100} = \frac{1}{3}$$

To change a percentage to a decimal numeral, remember that **To Decimal** the percent symbol means *divided by 100* and that to divide by **Numerals** 100 means to move the decimal point two places to the left (Section 2.5).

$$2\% = \frac{2}{100} \quad = 0.02$$

$$15\% = \frac{15}{100} \quad = 0.15$$

$$150\% = \frac{150}{100} \quad = 1.5$$

$$200\% = \frac{200}{100} \quad = 2$$

$$12\frac{1}{2}\% = 12.5\% = \frac{12.5}{100} = 0.125$$

$$33\frac{1}{3}\% = \frac{1}{3} \quad = 0.\overline{3}$$

Remember, when a percentage appears in a problem, before it can be used in calculations, it first must be converted to a common fraction or a decimal numeral.

Convert the following percentages to equivalent common fractions.   **Exercises 3.1** Reduce if necessary.

1. 18%   2. 250%

3. $8\frac{1}{2}\%$ 4. 0.3%

5. 450%   6. 6.19%

7. 2%   8. $12\frac{3}{4}\%$

9. 0.03%   10. 0.015%

Convert the following percentages to equivalent decimal numerals.

11. 4%   12. $15\frac{1}{5}\%$

13. 3.94%   14. 325%

15. $\frac{1}{8}\%$ 16. 50%

17. 100%   18. 0.5%

19. 0.03%   20. 0.015%

21. A drug is available in 0.1% solution NF and 0.02% solution nasal. Express each of these percentages as equivalent decimal numerals and as common fractions.

22. Propoxycaine hydrochloride NF is available in 0.4% solution. Express this percentage as an equivalent decimal numeral and common fraction.

23. Betamethasone benzoate is available in cream, 0.025%. Express this percentage as an equivalent decimal numeral and as a common fraction.

24. A drug is available in 0.125% solution with sodium bicarbonate 2% and glycerin 5%. Express each of these percentages as equivalent decimal numerals and common fractions.

Section 3.2   **Converting From Decimal Numerals and Common Fractions**

From Decimal Numerals   To change a decimal numeral to a percentage, read the decimal in terms of hundredths (Section 2.2). The number of hundredths gives the percentage.

$$0.16 = \frac{16}{100} = 16\%$$

$$0.165 = \frac{16.5}{100} = 16.5\%$$

$$1.56 = \frac{156}{100} = 156\%$$

$$0.\overline{3} = \frac{33.\overline{3}}{100} = 33.\overline{3}\%^*$$

---

*Or, $0.\overline{3} = \dfrac{33\frac{1}{3}}{100} = 33\frac{1}{3}\%$.

$$0.005 = \frac{0.5}{100} = 0.5\%$$

$$0.2 \ = 0.20 = \frac{20}{100} = 20\%$$

Therefore, *to change a decimal numeral to a percentage, move the decimal point two places to the right and attach a percent symbol.*

To change a common fraction to a percentage, two methods are useful. **From Common Fractions**

1. If it is easy to change the denominator to 100 do so, and use the numerator as the percentage.

$$\frac{1}{2} = \ \frac{1}{2} \cdot \frac{50}{50} = \frac{50}{100} = 50\%$$

$$\frac{1}{4} = \ \frac{1}{4} \cdot \frac{25}{25} = \frac{25}{100} = 25\%$$

$$\frac{3}{50} = \ \frac{3}{50} \cdot \frac{2}{2} = \frac{6}{100} = 6\%$$

$$\frac{75}{50} = \ \frac{75}{50} \cdot \frac{2}{2} = \frac{150}{100} = 150\%$$

$$\frac{7.5}{50} = \ \frac{7.5}{50} \cdot \frac{2}{2} = \frac{15}{100} = 15\%$$

$$\frac{0.75}{50} = \frac{0.75}{50} \cdot \frac{2}{2} = \frac{1.5}{100} = 1.5\%$$

2. In all other cases, first change the common fraction to a decimal numeral correct to the thousandth place (Section 2.7), and then change the decimal numeral to a percentage.

$$\frac{1}{8} = 0.125 = 12.5\%$$

$$\frac{2}{7} = 0.286 = 28.6\%$$

$$\frac{2}{3} = 0.667 = 66.7\%*$$

---

*Or, $\frac{2}{3} = 0.66\frac{2}{3} = 66\frac{2}{3}\%$.

$$\frac{150}{60} = 2.50 \ = 250\%$$

$$\frac{3}{80} = 0.038 = 3.8\%$$

$$\frac{3}{800} = 0.004 = 0.4\%*$$

Exercises 3.2   Convert to equivalent percentages.

1.	0.14	2.	0.03
3.	0.425	4.	0.132
5.	1	6.	12.025
7.	0.0003	8.	$0.\overline{6}$
9.	$\dfrac{2}{25}$	10.	$\dfrac{5}{4}$
11.	$\dfrac{7}{12}$	12.	$\dfrac{3}{8}$
13.	$\dfrac{1}{20}$	14.	$\dfrac{1}{1000}$
15.	$\dfrac{5}{6}$	16.	$\dfrac{3}{500}$

Section 3.3   **Operations Using Common Fractions, Decimal Numerals, and Percentages**

Addition and Subtraction   To perform the operations of addition or subtraction that involve mixtures of common fractions and decimal numerals, convert all the numerals either to common fractions or to decimal numerals. It does not matter which, although sometimes one form will be easier to work with than the other.

---

*Or, $\dfrac{3}{800} = 0.0038 = 0.38\%$.

Add $\frac{1}{2}$ and 0.6.

EXAMPLE 1

$$\frac{1}{2} + 0.6 = \left(\frac{5}{5} \cdot \frac{1}{2}\right) + \frac{6}{10} \qquad \text{(common fractions)}$$

$$= \frac{5}{10} + \frac{6}{10}$$

$$= \frac{11}{10}$$

$$= 1\frac{1}{10}$$

$$\frac{1}{2} + 0.6 = 0.5 + 0.6 \qquad \text{(decimal numerals)}$$

$$
\begin{array}{r}
0.5 \\
+0.6 \\
\hline
1.1
\end{array}
$$

Subtract $\frac{1}{3}$ from 0.8.

EXAMPLE 2

$$0.8 - \frac{1}{3} = \left(\frac{3}{3} \cdot \frac{8}{10}\right) - \left(\frac{1}{3} \cdot \frac{10}{10}\right) \qquad \text{(common fractions)}$$

$$= \frac{24}{30} - \frac{10}{30}$$

$$= \frac{14}{30}$$

$$= \frac{7}{15}$$

$$0.8 - \frac{1}{3} = 0.8 - 0.3* \qquad \text{(decimal numerals)}$$

$$
\begin{array}{r}
0.8 \\
-0.3 \\
\hline
0.5
\end{array}
$$

Because we round off when converting to decimals, 0.5 is slightly larger than $\frac{7}{15}$. Seven-fifteenths is the exact answer.

---

*$\frac{1}{3}$ rounded. 0.3 is slightly smaller than $\frac{1}{3}$.

Multiplication and Division

To perform the operations of multiplication and division that involve mixtures of common fractions and decimal numerals, you have the choice of working with the mixture or of converting to the same type of numeral.

EXAMPLE 3

Multiply $\frac{3}{4}$ by 1.96.

$$\frac{3}{4} \times 1.96 = \frac{3}{4} \times \frac{1.96}{1} \qquad \text{(mixture)}$$

$$= \frac{5.88}{4}$$

$$= 1.47$$

$$\frac{3}{4} \times 1.96 = \frac{3}{4} \times \frac{196}{100} \qquad \text{(common fractions)}$$

$$= \frac{147}{100}$$

$$= 1\frac{47}{100}$$

$$\frac{3}{4} \times 1.96 = 0.75 \times 1.96 \qquad \text{(decimal fractions)}$$

$$
\begin{array}{r}
1.96 \\
\times\, 0.75 \\
\hline
9\ 80 \\
1\ 37\ 2 \\
\hline
1.4700
\end{array}
$$

$$= 1.47$$

EXAMPLE 4

Divide 0.45 by $\frac{3}{2}$.

$$0.45 \div \frac{3}{2} = \frac{0.45}{1} \div \frac{3}{2} \qquad \text{(mixture)}$$

$$= \frac{0.45}{1} \cdot \frac{2}{3}$$

$$= \frac{0.90}{3}$$

$$= 0.30$$

EXAMPLE 4
continued

$$0.45 \div \frac{3}{2} = \frac{45}{100} \div \frac{3}{2} \qquad \text{(common fractions)}$$

$$= \frac{45}{100} \cdot \frac{2}{3}$$

$$= \frac{3}{10}$$

$$0.45 \div \frac{3}{2} = 0.45 \div 1.5 \qquad \text{(decimal fractions)}$$

$$= 15\overline{)4.5}$$

$$= 0.3$$

As noted earlier, when operations involve percentages, the percentages first must be converted to common fractions or to decimal numerals before the operations are performed. One type of percentage problem is considered here; others will be discussed in later chapters.

What is 15 percent of 30?

EXAMPLE 5

$$15\% \text{ of } 30 = \frac{15}{100} \cdot \frac{30}{1} \qquad \text{(common fractions)}$$

$$= \frac{450}{100}$$

$$= \frac{9}{2}$$

$$= 4\frac{1}{2}$$

$$15\% \text{ of } 30 = 0.15 \times 30 \qquad \text{(decimal fractions)}$$

$$= \begin{array}{r} 30 \\ \times 0.15 \\ \hline 1\ 50 \\ 3\ 0 \\ \hline 4.50 \end{array}$$

$$= 4.5$$

EXAMPLE 6	What is $33\frac{1}{3}$ percent of 120?

$$33\frac{1}{3}\% \text{ of } 120 = \frac{1}{3} \times \frac{120}{1} \qquad \text{(common fractions)}$$

$$= \frac{120}{3}$$

$$= 40$$

$$33\frac{1}{3}\% \text{ of } 120 = 0.333^* \times 120 \qquad \text{(decimal fractions)}$$

$$
\begin{array}{r}
120 \\
\times 0.333 \\
\hline
360 \\
3\,60 \\
36\,0 \\
\hline
39.960
\end{array}
$$

$$= 39.96$$

In Chapters 1 and 2 we noted that the verbal expressions *is what part of* and *is how many times* each indicate the operation of division (Section 1.7). Another verbal expression *is what percentage of* also *indicates the operation of division and the resulting common fraction or decimal numeral is expressed as a percent.*

EXAMPLE 7	8 is what percentage of 10?

This verbal expression translates as $8 \div 10$ or $\dfrac{8}{10}$ or 0.8

$$\frac{8}{10} = \frac{80}{100}$$

$$= 80\%$$

8 is 80% of 10

---

*We've rounded $33\frac{1}{3}\%$. 0.333 is slightly smaller than $33\frac{1}{3}\%$, so the answer is slightly smaller than 40.

$4\frac{1}{2}$ is what percentage of 25?

EXAMPLE 8

This verbal expression translates as $4\frac{1}{2} \div 25$ or $\dfrac{4\frac{1}{2}}{25}$ or $\dfrac{4.5}{25.0}$

$$\frac{4\frac{1}{2}}{25} = \frac{4\frac{1}{2}}{25} \cdot \frac{4}{4}$$

$$= \frac{18}{100}$$

$$= 18\%$$

$$\text{or } \frac{4.5}{25.0} = 0.18$$

$$= 18\%$$

$4\frac{1}{2}$ is 18% of 25

0.12 is what percentage of 3.5?

EXAMPLE 9

This verbal expression translates as $0.12 \div 3.5$ or $\dfrac{0.12}{3.50}$

$$\frac{0.12}{3.50} = \frac{12}{350}$$

$$= 0.034$$

$$= 3.4\%$$

0.12 is 3.4% of 3.5

6 is what percentage of 4?

EXAMPLE 10

This verbal expression translates as $6 \div 4$ or $\dfrac{6}{4}$

$$\frac{6}{4} = \frac{6}{4} \cdot \frac{25}{25}$$

$$= \frac{150}{100}$$

$$= 150\%$$

6 is 150% of 4

EXAMPLE 11

4 is what percentage of 6?

This verbal expression translates as $4 \div 6$ or $\dfrac{4}{6}$

$$\frac{4}{6} = \frac{2}{3}$$
$$= 0.67$$
$$= 67\%$$

4 is 67% of 6

Exercises 3.3   Perform the indicated operations.

1. $0.7 + \dfrac{3}{4} =$

2. $3.4 + 2\dfrac{3}{4} =$

3. $\dfrac{5}{6} + 1.2 =$

4. $2.1 + \dfrac{2}{7} =$

5. $\dfrac{1}{150} + 0.5 =$

6. $\dfrac{3}{4} - 0.7 =$

7. $3.4 - 2\dfrac{3}{4} =$

8. $1.2 - \dfrac{5}{6} =$

9. $2.1 - \dfrac{2}{7} =$

10. $0.5 - \dfrac{1}{150} =$

11. $2.1 \times \dfrac{2}{7} =$

12. $\dfrac{1}{2} \times 3.3 =$

13. $2\dfrac{3}{4} \times 6.5 =$

14. $\dfrac{1}{6} \times 0.5 =$

15. $0.002 \times \dfrac{3}{4} =$

16. $6.96 \div \dfrac{3}{2} =$

17. $6.96 \div \dfrac{5}{2} =$

18. $2\dfrac{3}{4} \div 0.05 =$

19. $1.8 \div 1\dfrac{1}{5} =$

20. $\dfrac{1}{150} \div 0.02 =$

21. 5% of $60 =$

22. 150% of $3.56 =$

23. $2\dfrac{1}{4}\%$ of $56 =$

24. 100% of $546\dfrac{2}{3} =$

25. 99.44% of 78 =   26. $12\frac{1}{2}$% of 96 =

27. $\frac{1}{6}$% of 0.12 =   28. 8.3% of 2.4 =

29. 0.7% of 0.7 =   30. $\frac{1}{4}$% of $\frac{4}{5}$ =

31. $\frac{3}{5}$ is what part of 3.6?   32. $4\frac{1}{3}$ is how many times 1.3?

33. $\frac{3}{5}$ is what percentage of   34. 4.5 is what percentage of
     3.6?   1.5?

[3.1]   **Review Exercises**

1. Write the following percentages as equivalent common frac-
    tions. Reduce if necessary.

    a.   5%   b. $66\frac{2}{3}$%

    c.   315%   d.   0.008%

2. Write the following percentages as equivalent decimal nu-
    merals.

    a.   5%   b. $66\frac{2}{3}$%

    c.   315%   d.   0.008%

3. Thimersol in concentrations ranging from 0.0025% to 0.1% is
    used for wet dressings. Express each (0.0025% and 0.1%) as a
    common fraction and as a decimal numeral.

[3.2]

4. Change the following to equivalent percentages.

    a.   0.15   b.   1.5

    c. $0.00\overline{6}$ d. $\frac{17}{25}$

    e. $\frac{1}{5000}$ f. $\frac{5}{3}$

[3.3]

5. Perform the indicated operations.

    a. $0.08 + \frac{3}{4} =$ b. $1.35 - \frac{7}{8} =$

c. $\dfrac{1}{6} \times 0.102 =$

d. $\dfrac{1}{300} \div 0.06 =$

e. $33\dfrac{1}{3}\%$ of $85 =$

f. $\dfrac{3}{4}\% \div \dfrac{3}{4} =$

g. $15\% + 1.5 =$

h. 0.2% of $12.5 =$

6. $\frac{1}{8}$ is what part of 2? $\frac{1}{8}$ is what percentage of 2?

7. $\frac{1}{8}$ is how many times 0.05? $\frac{1}{8}$ is what percentage of 0.05?

8. Place $=, >,$ or $<$ between each pair of numerals to make a true statement.

a. 1.5% 0.015

b. $\dfrac{7}{8}$ 87%

c. $5\dfrac{1}{2}\%$ $\dfrac{11}{2}$

d. 3.7% 3.7

Ratios and Proportions

In health-related occupations the majority of mathematical problems are ratio and proportion in form. This chapter introduces ratios and proportions and explains their applications to medication problems. The first medical problems are of the simplest type, and frequently you can get the answers in your head. This is intentional so that you can focus your attention on the logic involved, the most error-free working form when setting up a problem, the correct method of solution, and methods for checking.

Topics include:

Definition of ratio, and setting up ratios from verbal statements

Definition of proportion, and solution and checking forms

Forms for setting up, labeling, working, and checking medication problems

Two types of medication problems:
> calculation of desired medication per dose
> calculation of total dosage over a period of time given certain patient characteristics

Medication problems that distinguish method of drug administration—that is, PO as a solid, PO as a liquid, IM or SC of an injectible liquid, suppositories

Section 4.1   Ratios

A *ratio* is the comparison of two quantities a and b, commonly written $a:b$ or $\frac{a}{b}$. a and b may be any numbers, except b may not be zero. For example, the ratio of 3 to 4 is

$$3:4, \text{ or } \frac{3}{4}$$

The ratio of $7\frac{1}{2}$ to 15 is

$$7\frac{1}{2}:15, \text{ or } \frac{7\frac{1}{2}}{15}, \text{ or } \frac{1}{2}$$

The ratio of 0.1 to 0.03 is

$$0.1:0.03, \text{ or } \frac{0.1}{0.03}, \text{ or } \frac{10}{3}$$

Note that a ratio may be a complex fraction $\left(\frac{7\frac{1}{2}}{15}\right)$ or be reduced to a simplified fraction $\left(\frac{1}{2}\right)$; a ratio may be a decimal fraction $\left(\frac{0.1}{0.03}\right)$ or be simplified to an improper fraction $\left(\frac{10}{3}\right)$. Any of these forms is an acceptable working form. However, you cannot convert an improper fraction to a mixed number: $\frac{10}{3}$ is a ratio; $3\frac{1}{3}$ is not a ratio.

When writing a ratio, if you have enough information to label the two quantities being compared, then you must label.

EXAMPLE 1

There are 8 ounces in 1 cup. Express this relationship in ratio form.

$$8 \text{ ounces}:1 \text{ cup}$$

or

$$\frac{8 \text{ ounces}}{1 \text{ cup}}$$

EXAMPLE 2

1 cup contains 8 ounces. Express this relationship in ratio form.

$$1 \text{ cup}:8 \text{ ounces}$$

or

$$\frac{1 \text{ cup}}{8 \text{ ounces}}$$

EXAMPLE 2
continued

Examples 1 and 2 deal with the same concept. When labels are used, the ratio 1 cup : 8 ounces has the same meaning as the ratio 8 ounces : 1 cup.

$$\frac{1 \text{ cup}}{8 \text{ ounces}} \text{ means } \textit{1 cup contains 8 ounces}$$

$$\frac{8 \text{ ounces}}{1 \text{ cup}} \text{ means } \textit{there are 8 ounces in 1 cup}$$

One aspect of solving medication problems entails setting up labeled ratios—translating the problem into ratio form.

1 litre solution contains 10 grams drug.

$$1 \text{ litre} : 10 \text{ grams or } \frac{1 \text{ litre}}{10 \text{ grams}}$$

EXAMPLE 3

or

$$10 \text{ grams} : 1 \text{ litre or } \frac{10 \text{ grams}}{1 \text{ litre}}$$

A bottle of 5 grain aspirin tablets.*

$$5 \text{ grains} : 1 \text{ tablet or } \frac{5 \text{ grains aspirin}}{1 \text{ tablet}}$$

$$1 \text{ tablet} : 5 \text{ grains or } \frac{1 \text{ tablet}}{5 \text{ grains aspirin}}$$

EXAMPLE 4

---

*Proprietary (brand or trade) names are indicated by a capital letter. Nonproprietary (generic or public) names are indicated by a lowercase letter.

EXAMPLE 5

Polymyxin B sulfate solution, otic, is available 10 000 units per millilitre.

$$10\ 000 \text{ units} : 1 \text{ millilitre}$$

or

$$\frac{10\ 000 \text{ units polymyxin}}{1 \text{ millilitre}}$$

EXAMPLE 6

Calciferol is available 1.25 milligrams per capsule.

$$1.25 \text{ milligrams} : 1 \text{ capsule}$$

or

$$\frac{1.25 \text{ milligrams calciferol}}{1 \text{ capsule}}$$

EXAMPLE 7

20 milligrams drug per kilogram body mass

$$20 \text{ milligrams} : 1 \text{ kilogram}$$

or

$$\frac{20 \text{ milligrams}}{1 \text{ kilogram}}$$

EXAMPLE 8

3% hydrogen peroxide solution (Section 3.1).

$$3 \text{ parts hydrogen peroxide} : 100 \text{ parts whole solution}$$

or

$$\frac{3 \text{ parts hydrogen peroxide}}{100 \text{ parts whole solution}}$$

Exercises 4.1

In each of the following exercises, translate the words into two different ratio forms. For example, *administer penicillin 25 000 units per kilogram body mass* becomes

$$\frac{25\ 000 \text{ units penicillin}}{1 \text{ kilogram body mass}} \text{ or } 25\ 000 \text{ units} : 1 \text{ kilogram}$$

1. The ratio of $\frac{1}{4}$ to $\frac{7}{8}$.

2. Amoxicillin is available as 250 milligram capsules.

3. Sodium citrate is available as 500 milligram capsules.

4. Scopolamine is available as 400 microgram, and 600 microgram.

5. A drug is available 50 000 units per tablet.

6. Bacitracin may be mixed to provide 5000 units bacitracin per millilitre.

7. Regular Iletin is available 100 units per millilitre in 10 millilitre vials.

8. Scopolamine hydrobromide is available for injection 300 micrograms per millilitre is 20 millilitre vials.

9. A drug is available 100 milligrams per millilitre in 10 millilitre ampuls.

10. The syrup is available 1.2 grams per 5 millilitres.

11. 10 grams drug dissolved in distilled water to make 150 millilitres solution.

12. 1 litre solution contains two $3\frac{1}{2}$ grain tablets.

13. 1 cup sugar dissolved in water to make a quart of solution.

14. Administer phenobarbital 0.125 micrograms per square metre body surface.

15. Give Dimethane 15 milligrams per square metre body surface.

16. Administer 20 millilitres 6% dextran per kilogram body mass.

17. Give Choloxin 50 micrograms per kilogram body mass.

18. Over a period of 6 hours, administer 300 milligrams drug.

19. Administer intravenously 2 litres solution over a period of 18 hours.

20. Administer intravenously 200 millilitres 5% hypertonic saline over a period of 4 hours.

21. Administer 1200 millilitres of 6% dextran during 24 hours.

22. 5% boric acid solution.

23. 5% sodium chloride solution.

Proportions

A *proportion* is defined as two equal ratios. For example, $\frac{3}{4}$ and $\frac{6}{8}$ are equal ratios, and they form a proportion that is written

$$\frac{3}{4} = \frac{6}{8}$$

Other examples of proportions are

$$\frac{1}{2} = \frac{5}{10} \qquad \frac{\frac{1}{3}}{12} = \frac{\frac{1}{4}}{9}$$

Recall that if two fractions (ratios) are equal, their cross products are equal (Section 1.9). By definition, then, the cross products of a proportion are equal.

If

$$\frac{3}{4} = \frac{6}{8} \qquad \text{then } 3 \times 8 = 4 \times 6$$

if

$$\frac{1}{2} = \frac{5}{10} \qquad \text{then } 1 \times 10 = 2 \times 5$$

if

$$\frac{\frac{1}{3}}{12} = \frac{\frac{1}{4}}{9} \qquad \text{then } \frac{1}{3} \times 9 = \frac{1}{4} \times 12$$

The cross product check together with the Axioms of Equality are used to solve proportions.

The Axioms of Equality

1. The sum of equalities is an equality. Or, if the same number is added to each member of an equality, the result is an equality.

    If $a = b$ and $c = c$ then $a + c = b + c$.

2. The difference of equalities is an equality. Or, if the same number is subtracted from each member of an equality, the result is an equality.

   If $a = b$ and $c = c$ then $a - c = b - c$.

3. The product of equalities is an equality. Or, if each member of an equality is multiplied by the same number, the result is an equality.

   If $a = b$ and $c = c$ then $a \cdot c = b \cdot c$.

4. The quotient of equalities is an equality. Or, if each member of an equality is divided by the same number, the result is an equality.

   If $a = b$ and $c = c$ then $\dfrac{a}{c} = \dfrac{b}{c}$.

To solve a proportion when one position of the proportion is unknown, let the unknown position be represented by x. The proportion is solved when the left member contains one x and all the numbers are in the right member.

Solving Proportions

$$\frac{3}{4} = \frac{x}{8}$$

EXAMPLE 1

Read *3 is to 4 as x is to 8.* If two fractions are equal their cross products also are equal:

$$4 \cdot x = 8 \cdot 3$$
$$4x = 24$$

To obtain $1x$ in the left member, apply Axiom 4 and divide both members by 4:

$$\frac{4 \cdot x}{4} = \frac{24}{4}$$
$$1x = 6$$
$$x = 6$$

EXAMPLE 2

$$\frac{3}{\frac{1}{2}} = \frac{x}{1\frac{1}{2}}$$

Read *3 is to $\frac{1}{2}$ as x is to $1\frac{1}{2}$*. In a proportion the cross products are equal:

$$\frac{1}{2} \cdot x = 1\frac{1}{2} \cdot 3$$

$$\frac{1}{2}x = \frac{9}{2}$$

Applying Axiom 3, multiply each member by 2 so that $1x$ is all that remains in the left member.

$$2 \cdot \frac{1}{2}x = \frac{9}{2} \cdot 2$$

$$1x = 9$$

$$x = 9$$

It is easiest first to compute the cross product that contains the x, putting the x in the left member.

EXAMPLE 3

$$\frac{0.05}{100} = \frac{0.025}{x}$$

$$(0.05)(x) = (100)(0.025)$$

$$0.05x = 2.5$$

$$\frac{0.05x}{0.05} = \frac{2.5}{0.05}$$

$$1x = 50$$

$$x = 50$$

A proportion may be written in ratio form. If so, first translate each ratio to common fraction form and then solve.

EXAMPLE 4

$$x : 3 = 6 : 24$$

$$\frac{x}{3} = \frac{6}{24}$$

$$24 \cdot x = 3 \cdot 6$$

EXAMPLE 4
continued

$$24x = 18$$

$$\frac{24x}{24} = \frac{18}{24}$$

$$1x = \frac{3}{4}$$

$$x = \frac{3}{4}$$

EXAMPLE 5

$$125 : x = 2.2 : 1$$

Read *125 is to x as 2.2 is to 1*. Translate each ratio to common fraction form.

$$\frac{125}{x} = \frac{2.2}{1}$$

Find the cross products.

$$(2.2)(x) = (1)(125)$$
$$2.2x = 125$$

Divide each member of the equation by 2.2.

$$\frac{2.2x}{2.2} = \frac{125.}{2.2}$$

$$1x = 56.8$$

$$x = 56.8$$

The left member contains one x, and all other numbers are in the right member. The equation is solved.

Checking

It is important to check the solution of a proportion. Two checking methods are suggested here. In both methods the first step is the same.

Cross product check. Substitute your answer for x in the original problem. Then cross multiply to see if the fractions are equal. If they are, the solution is correct.

Reduction check. Substitute your answer for x in the original problem. Simplify both the left and right members to see if the two members are identical. If they are, the solution is correct.

Using both methods, check the five preceding examples.

**EXAMPLE 1
CHECK**

$$\frac{3}{4} \overset{?}{=} \frac{6}{8}$$ (cross product check)

$$3 \cdot 8 \overset{?}{=} 4 \cdot 6$$

$$24 = 24 \quad \checkmark$$

$$\frac{3}{4} \overset{?}{=} \frac{6}{8}$$ (reduction check)

$$\frac{3}{4} \overset{?}{=} \frac{2 \cdot 3}{2 \cdot 2 \cdot 2}$$

$$\frac{3}{4} \overset{?}{=} \frac{\cancel{2} \cdot 3}{\cancel{2} \cdot 2 \cdot 2}$$

$$\frac{3}{4} = \frac{3}{4} \quad \checkmark$$

**EXAMPLE 2
CHECK**

$$\frac{3}{\frac{1}{2}} \overset{?}{=} \frac{9}{1\frac{1}{2}}$$ (cross product check)

$$3 \cdot 1\frac{1}{2} \overset{?}{=} 9 \cdot \frac{1}{2}$$

$$\frac{9}{2} = \frac{9}{2} \quad \checkmark$$

$$\frac{3}{\frac{1}{2}} \overset{?}{=} \frac{9}{1\frac{1}{2}}$$ (reduction check)

$$3 \div \frac{1}{2} \overset{?}{=} 9 \div \frac{3}{2}$$

$$\frac{3}{1} \cdot \frac{2}{1} \overset{?}{=} \frac{9}{1} \cdot \frac{2}{3}$$

$$\frac{6}{1} \overset{?}{=} \frac{3 \cdot 3 \cdot 2}{3}$$

$$\frac{6}{1} \overset{?}{=} \frac{\cancel{3} \cdot 3 \cdot 2}{\cancel{3}}$$

$$6 = 6 \quad \checkmark$$

$$\frac{0.05}{100} \stackrel{?}{=} \frac{0.025}{50}$$ (cross product check)

$$(50)(0.05) \stackrel{?}{=} (100)(0.025)$$

$$2.5 = 2.5 \quad \checkmark$$

$$\frac{0.05}{100} \stackrel{?}{=} \frac{0.025}{50}$$ (reduction check)

$$0.0005 = 0.0005 \quad \checkmark$$

EXAMPLE 3 CHECK

$$\frac{\frac{3}{4}}{3} \stackrel{?}{=} \frac{6}{24}$$ (cross product check)

$$\frac{3}{4} \cdot 24 \stackrel{?}{=} 3 \cdot 6$$

$$18 = 18 \quad \checkmark$$

$$\frac{\frac{3}{4}}{3} \stackrel{?}{=} \frac{6}{24}$$ (reduction check)

$$\frac{3}{4} \div 3 \stackrel{?}{=} \frac{6}{24}$$

$$\frac{3}{4} \cdot \frac{1}{3} \stackrel{?}{=} \frac{6}{24}$$

$$\frac{1}{4} = \frac{1}{4} \quad \checkmark$$

EXAMPLE 4 CHECK

$$\frac{125}{56.8} \stackrel{?}{=} \frac{2.2}{1}$$ (cross product check)

$$(125)(1) \stackrel{?}{=} (56.8)(2.2)$$

$$125 \approx 124.96 \quad \checkmark$$

$$\frac{125}{56.8} \stackrel{?}{=} \frac{2.2}{1}$$ (reduction check)

$$2.2007 \approx 2.2 \quad \checkmark$$

EXAMPLE 5 CHECK

When a solution has been rounded off, as it was in Example 5, the checks will be approximate checks. ($\approx$ means *approximately equal to*.)

Exercises 4.2   Solve and check.

1. $3:2 = x:6$

2. $3:2 = 6:x$

3. $0.5:x = 1:2$

4. $1:2 = \dfrac{1}{2}:x$

5. $x:27 = 3:5$

6. $27:x = 3:5$

7. $5:3 = x:27$

8. $5:3 = 27:x$

9. $5:7 = 25:x$

10. $7:5 = x:25$

11. $1:2.2 = 15:x$

12. $2.2:1 = x:15$

13. $0.025:x = 0.06:1$

14. $0.025:x = 6:100$

15. $x:0.12 = 100:3$

16. $\dfrac{0.6}{x} = \dfrac{3.6}{2}$

17. $\dfrac{x}{0.3} = \dfrac{1.5}{0.9}$

18. $\dfrac{6}{100} = \dfrac{x}{\dfrac{1}{4}}$

19. $\dfrac{3}{5} = \dfrac{4\dfrac{1}{2}}{x}$

20. $\dfrac{x}{2\dfrac{3}{4}} = \dfrac{\dfrac{1}{2}}{\dfrac{3}{8}}$

21. $\dfrac{\dfrac{1}{7}}{\dfrac{1}{4}} = \dfrac{x}{\dfrac{1}{16}}$

22. $\dfrac{\dfrac{1}{150}}{x} = \dfrac{\dfrac{1}{200}}{\dfrac{1}{300}}$

23. $\dfrac{\dfrac{1}{2}}{2\dfrac{1}{2}} = \dfrac{2}{x}$

24. $\dfrac{x}{2} = \dfrac{30}{\dfrac{1}{2}}$

25. $\dfrac{1.2}{3.5} = \dfrac{10}{x}$

26. $\dfrac{6}{100} = \dfrac{x}{100}$

27. $\dfrac{0.03}{x} = \dfrac{0.5}{10}$

28. $\dfrac{60}{1} = \dfrac{750}{x}$

29. $\dfrac{x}{4000} = \dfrac{0.001}{1}$

30. $\dfrac{1}{2.2} = \dfrac{x}{125}$

31. $\dfrac{1}{2.2} = \dfrac{x}{80}$ 32. $\dfrac{2}{x} = \dfrac{1}{65}$

Medication Problems   Section 4.3

Ratios and proportions are used extensively in the solution of health-related problems. One method of solution is as follows:

1. Translate the given information into a proportion (Section 4.1).
2. It is mandatory when sufficient information is available that each position of the proportion be labeled.
3. In the type of proportion illustrated in this section the pattern of the position of the labels must be the same in both the left member and the right member. In any type of proportion the labeling must be consistent.
4. Solve the proportion, being sure to label the answer.
5. Check the solution.

If a cup contains 8 ounces, how many ounces are there in $1\frac{3}{4}$ cups? To begin, translate the given information into a proportion:

EXAMPLE 1

$$1 \text{ cup contains 8 ounces} = \frac{1 \text{ cup}}{8 \text{ ounces}}$$

$$\text{how many ounces in } 1\tfrac{3}{4} \text{ cups} = \frac{1\dfrac{3}{4} \text{ cups}}{x \text{ ounces}}$$

So the proportion becomes

$$\frac{1 \text{ cup}}{8 \text{ ounces}} = \frac{1\dfrac{3}{4} \text{ cups}}{x \text{ ounces}}$$

EXAMPLE 1
continued

Note that the left member is cup : ounces; therefore, the right member must follow the same pattern (cup : ounces). Now, solve the proportion:

$$1 \cdot x = 8 \cdot 1\frac{3}{4}$$ (cross product
$$x = 14 \text{ ounces}$$ equality)

Label the answer *ounces* because the x was labeled *ounces* in the original proportion. There are 14 ounces in $1\frac{3}{4}$ cups. Check the solution:

$$\frac{1}{8} \overset{?}{=} \frac{1\frac{3}{4}}{14}$$ (reduction check)

$$\frac{1}{8} \overset{?}{=} 1\frac{3}{4} \div 14$$

$$\frac{1}{8} \overset{?}{=} \frac{7}{4} \cdot \frac{1}{14}$$

$$\frac{1}{8} = \frac{1}{8} \quad \checkmark$$

Note that no labels are required in the check. The check is to see if your calculations are correct.

Now, let's apply the procedure to several types of medication problems:

---

1. Oral administration (PO) as a solid, or in tablet or capsule form
2. Oral administration (PO) as a liquid, or in syrup, elixir, or solution form
3. Intramuscular (IM or i.m.) or subcutaneous (SC or s.c.) administration of an injectable liquid
4. Rectal or vaginal insertion of suppositories

A fifth method (intravenous administration) is discussed in Chapter 10. In addition to the four types of problems just noted, illustrated in this chapter are two methods for determining the total amount of medication desired for a patient: quantity of drug per kilogram body mass (weight) and quantity of drug per square metre body surface.

> The *medication on hand* is the drug as it is available in the supply cupboard or from the pharmacist. The *desired medication* is the amount of the drug that the doctor has ordered to be given to the patient.
>
> Medication-on-hand ratio = Desired-medication ratio

The doctor asks you to administer 2.5 milligrams calciferol to your patient. 1.25 milligram calciferol capsules are available.

medication on hand		desired medication

$$\frac{1.25 \text{ milligrams}}{1 \text{ capsule}} = \frac{2.5 \text{ milligrams}}{x \text{ capsules}}$$

$$1.25x = 2.5$$

$$\frac{1.25x}{1.25} = \frac{2.5}{1.25}$$

$$x = 2 \text{ capsules calciferol}$$
$$(1.25 \text{ milligrams/tab})$$

$$\frac{1.25}{1} \overset{?}{=} \frac{2.5}{2} \qquad \text{(reduction check)}$$

$$1.25 = 1.25 \quad \checkmark$$

**EXAMPLE 2:
ORAL ADMIN-
ISTRATION,
SOLIDS**

The doctor asks you to administer 250 milligrams simethicone to your patient. Simethicone 40 milligrams per 0.6 millilitre is available.

medication on hand		desired medication

$$\frac{40 \text{ milligrams}}{0.6 \text{ millilitre}} = \frac{250 \text{ milligrams}}{x \text{ millilitres}}$$

**EXAMPLE 3:
ORAL ADMIN-
ISTRATION,
LIQUIDS**

EXAMPLE 3:
ORAL ADMIN-
ISTRATION,
LIQUIDS
continued

$$40x = 150$$

$$\frac{40x}{40} = \frac{150}{40}$$

$$x = 3.75 \text{ millilitres}$$
simethicone
(40 milligrams/0.6 millilitres)

$$\frac{40}{0.6} \overset{?}{=} \frac{250}{3.75} \qquad \text{(cross product check)}$$

$$(40)(3.75) \overset{?}{=} (250)(0.6)$$

$$150 = 150 \quad \checkmark$$

EXAMPLE 4:
INJECTIONS,
IM AND SC

The doctor asks you to administer atropine sulfate 220 micrograms. There is on hand atropine sulfate solution 400 micrograms per millilitre. How many millilitres will you administer?

$$\underset{\substack{\text{medication} \\ \text{on hand}}}{\frac{400 \text{ micrograms}}{1 \text{ millilitre}}} = \underset{\substack{\text{desired} \\ \text{medication}}}{\frac{220 \text{ micrograms}}{x \text{ millilitres}}}$$

$$400x = 220$$

$$\frac{400x}{400} = \frac{220}{400}$$

$$x = 0.55 \text{ millilitre atropine sulfate}$$
(400 micrograms/millilitre)

$$\frac{400}{1} \overset{?}{=} \frac{220.}{0.55} \qquad \text{(cross product check)}$$

$$(400)(0.55) \overset{?}{=} (220)(1)$$

$$220 = 220 \quad \checkmark$$

EXAMPLE 5:
SUPPOSI-
TORIES

Theophylline 125 milligram suppositories are available. The doctor asks you to administer 250 milligrams during 24 hours. How many suppositories are needed?

$$\underset{\substack{\text{medication} \\ \text{on hand}}}{\frac{125 \text{ milligrams}}{1 \text{ suppository}}} = \underset{\substack{\text{desired} \\ \text{medication}}}{\frac{250 \text{ milligrams}}{x \text{ suppositories}}}$$

EXAMPLE 5:
SUPPOSI-
TORIES
continued

$$125x = 250$$

$$\frac{125x}{125} = \frac{250}{125}$$

$x = 2$ **suppositories theophylline**
(125 milligrams/tab)

$$\frac{125}{1} \overset{?}{=} \frac{250}{2} \qquad \text{(reduction check)}$$

$$125 = 125 \quad \checkmark$$

The doctor asks you to administer 0.2 milligram drug per kilogram body mass. Your patient weighs 28 kilograms.

**EXAMPLE 6:
DRUG PER
KILOGRAM
BODY MASS**

$$\underset{\substack{\text{medication} \\ \text{ratio}}}{\frac{0.2 \text{ milligram}}{1 \text{ kilogram}}} = \underset{\text{dosage}}{\frac{x \text{ milligrams}}{28 \text{ kilograms}}}$$

$$x = (0.2)(28)$$

$$x = 5.6 \text{ milligrams drug}$$

$$\frac{0.2}{1} \overset{?}{=} \frac{5.6}{28} \qquad \text{(reduction check)}$$

$$0.2 = 0.2 \quad \checkmark$$

The doctor asks you to administer meprobamate 0.7 gram per square metre body surface. The patient has a body surface of 0.6 square metre.

**EXAMPLE 7:
DRUG PER
SQUARE
METRE BODY
SURFACE**

$$\underset{\substack{\text{medication} \\ \text{ratio}}}{\frac{0.7 \text{ gram}}{1 \text{ square metre}}} = \underset{\text{dosage}}{\frac{x \text{ gram}}{0.6 \text{ square metre}}}$$

$$x = (0.7)(0.6)$$

$$x = 0.42 \text{ gram meprobamate}$$

$$\frac{0.7}{1} \overset{?}{=} \frac{0.42}{0.6} \qquad \text{(reduction check)}$$

$$0.7 = 0.7 \quad \checkmark$$

EXAMPLE 8:
TABLETS
AVAILABLE

In Example 7 the dosage was calculated. Meprobamate tablets 0.2 gram are available. How many will you administer?

$$\frac{\overset{\text{medication}}{\underset{\text{on hand}}{}}}{} \qquad \frac{\overset{\text{desired}}{\underset{\text{medication}}{}}}{}$$

$$\frac{0.2 \text{ gram}}{1 \text{ tablet}} = \frac{0.4 \text{ gram}}{x \text{ tablets}}$$

$$0.2x = 0.4$$

$$\frac{0.2x}{0.2} = \frac{0.4}{0.2}$$

$$x = 2 \text{ tablets meprobamate}$$
$$(0.2 \text{ gram/tab})$$

$$\frac{0.2}{1} \overset{?}{=} \frac{0.4}{2} \qquad\qquad \text{(reduction check)}$$

$$0.2 = 0.2 \ \checkmark$$

Exercises 4.3   Solve for the information requested. Be sure to label and check your answers.

1. The doctor asks you to administer 1500 milligrams sodium bicarbonate. Sodium bicarbonate tablets 600 milligrams are available. How many will you give?

2. Methantheline bromide 50 milligram scored tablets are available. The doctor asks you to administer 25 milligrams. How many tablets will you give?

3. Desired medication: dicumarol 150 milligrams; medication on hand: dicumarol 100 milligram tablets. How many tablets will you administer?

4. The doctor asks you to administer benzonatate 300 milligrams. On hand are benzonatate capsules 100 milligrams. How many capsules will you give?

5. Desired medication: 500 milligrams; medication on hand: 750 milligram scored tablets. How many will you give?

6. Desired medication: nicotinyl alcohol 125 milligrams; medication on hand: nicotinyl alcohol 50 milligram scored tablets. How many tablets will you give?

7. Desired medication: Antrenyl Bromide 2.5 milligrams; medication on hand: Antrenyl Bromide 5 milligram scored tablets. How many tablets will you give?

8. Desired medication: ferrous sulfate 600 milligrams; medication on hand: ferrous sulfate 200 milligram tablets. How many tablets will you give?

9. Amobarbital elixir is available 44 milligrams per 5 millilitres solution. The doctor asks you to administer 50 milligrams amobarbital, PO. How much will you give?

10. Artane solution 2 milligrams per 5 millilitres is available. The doctor asks you to give 5 milligrams Artane. How many millilitres will you give?

11. Desired medication: histamine phosphate 150 micrograms; medication on hand: histamine phosphate 275 micrograms per millilitre. How much solution will you administer?

12. Desired medication: reserpine 3 milligrams; medication on hand: reserpine 2.5 milligrams per millilitre. How much will you give?

13. Desired medication: ferrous sulfate 375 milligrams; medication on hand: ferrous sulfate 125 milligrams per millilitre in calibrated dropper bottle. How much will you give?

14. The doctor asks you to administer 50 milligrams Dramamine. Dramamine 15.0 milligrams per millilitre is available. How much will you administer?

15. Testosterone propionate 100 milligrams per millilitre in 10 millilitre vials is available. The doctor asks you to give 275 milligrams. How much will you administer?

16. Estradiol 20 milligrams per millilitre in 5 millilitre vials is available. You are to give 15 milligrams estradiol. How much will you administer?

17. Frequently the dosage of a drug depends on the mass or weight of the patient. Scopolamine hydrobromide is administered 6 micrograms per kilogram body mass.

    a. How much will you give a 42 kilogram patient?

    b. Scopolamine hydrobromide is available as an injectable liquid 300 micrograms per millilitre. How many millilitres will you administer?

18. Chlor-Trimeton is administered 0.35 milligram per kilogram body mass total daily as 4 equally divided doses.

    a. How much will you give a 25 kilogram child?

    b. Chlor-Trimeton oral liquid is available in strength 2 milligrams per 5 millilitres. How many millilitres will you administer?

19. A total daily digitalizing dose of digoxin for a nine-year-old child is 15 micrograms per kilogram body mass.

    a. How much will you administer to a 34 kilogram patient?

    b. Digoxin elixir is available 50 micrograms per millilitre in a calibrated dropper bottle. How much will you administer total daily?

20. The desired dose of a drug may be based on the body surface in square metres of the patient. The daily dose of carbarsone is 300 milligrams per square metre body surface.

    a. How much carbarsone is needed for a patient with 0.84 square metre body surface?

    b. Carbarsone 250 milligram tablets are available. How many will you administer?

21. The 24-hour dose of theophylline is 0.3 gram per square metre body surface.

    a. How much theophylline will you administer in 24 hours to a patient with 1.1 square metres body surface?

    b. Theophylline elixir is available 0.027 gram per 5 millilitres. How much will you administer?

22. Calcium Lactate is prescribed 1.2 grams per square metre body surface over 24 hours.

    a. How much Calcium Lactate is needed for a patient with 0.76 square metre body surface?

    b. Calcium Lactate 1 gram wafers are available. How many wafers should be administered in 24 hours?

23. The doctor asks you to administer secobarbital 180 milligrams in 3 divided doses during 24 hours. Secobarbital 60 milligram suppositories are available.

    a. How many will you need over 24 hours?

    b. How many will you need per dose?

24. Bisacodyl suppositories 10 milligrams are available. The doc-
    tor asks you to administer 5 milligrams to an infant. How
    many suppositories will you need?

25. The doctor wants the administration of chloral hydrate 0.9
    gram per 24 hours given as three separate doses. Chloral
    hydrate suppositories 0.3 gram are available. How many will
    you need per dose?

26. Phenergan is prescribed 0.5 milligrams per kilogram body
    mass. The patient's mass is 24 kilograms; Phenergan supposi-
    tories 25 milligrams are available. How many will you need?

[4.1]   **Review Exercises**

Translate the following into two different ratio forms:

1. The ratio of $4\frac{1}{2}$ to 9
2. One (1) litre of solution contains two 3.5 gram tablets
3. Bupivacaine is available 0.25%

[4.2]
Solve and check these proportions:

4. $5:\dfrac{3}{4}=x:\dfrac{9}{8}$

5. $\dfrac{0.25}{x}=\dfrac{0.005}{0.1}$

[4.3]
Solve these medication problems. State the formula used. Be sure
to label where necessary, and check your answer.

6. Desired medication: fluphenazine 5 milligram
   Medication on hand: fluphenazine 2.5 milligram per 5 millilitre

7. A dosage of dimenhydrinate 1.25 milligram per kilogram body
   mass is desired.

   a. How much dimenhydrinate is required for a 30 kilogram
      child?

   b. Calculate the IM dosage if dimenhydrinate is available as
      50 milligrams per millilitre.

8. The desired dosage of erythromycin is 1600 milligrams per
   square metre body surface total daily as 4 equally divided
   doses. The patient has a body surface of 1.25 square metre.
   Erythromycin is available as 500 milligram tablets.

The Apothecaries' System of Measurement

This book explains various systems of measurement used in health-related occupations: apothecaries', metric (SI), household, U.S. Customary, and drugs measured in units. Two of these measurement systems are used only to measure the quantity of a substance: household and U.S. Customary; one is used only to measure strength of medication: drugs measured in units; and two are used to measure both quantity and strength of a substance: apothecaries' and metric (SI). This chapter introduces the apothecaries' system of measurement. The apothecaries' system is an ancient system not used in our more modern hospitals; however, who can say that each trained health professional will have lifetime employment in a "modern hospital"?

Topics include:

Roman numerals commonly used in medicine

Definitions, symbols, and values for mass and volume apothecaries' units of measurement, and a simple method for changing from one unit to another

Relationship between mass and volume units

Medication problems involving apothecaries' units of measurement

Roman Numerals

The profession of medicine is an ancient one, a fact reflected in its retention of some ancient language and symbolism. The use of Roman numerals is an example of this. Traditionally the only Roman numerals used are the smaller ones (i through xxx). A variation implemented by the health-related occupations is that lowercase letters, instead of capital letters, are used.

Arabic numerals	Roman numerals	Arabic numerals	Roman numerals
½	ss s̄s̄	4½	ivss ivs̄s̄ īvs̄s̄
1	i ī	5	v v̄
1½	iss īs̄s̄ īs̄s̄	5½	vss v̄s̄s̄
2	ii īi	6	vi v̄i
2½	iiss iīs̄s̄ īis̄s̄	7	vii v̄ii
3	iii īii	8	viii v̄iii
3½	iiiss iiīs̄s̄ īiis̄s̄	9	ix īx
4	iv īv	10	x x̄

Table 5.1

Table 5.1 shows Roman numerals $\frac{1}{2}$ to 10. To form Roman numerals 11 through 19, combine x (the symbol for 10) with i through ix (the symbols for 1 through 9) in this manner:

xi   xii   xiii   xiv   xv   xvi   xvii   xviii   xix

The symbol for 20 is xx. To form Roman numerals for 21 through 29, combine xx (the symbol for 20) with i through ix in this manner:

xxi   xxii   xxiii   xxiv   xxv   xxvi   xxvii   xxviii   xxix

The symbol for 30 is xxx.

Mass and Volume

The apothecaries' system is an ancient system of measurement. It has been used by the medical profession for hundreds of years, and although currently it is being phased out in the United States and in most of the world, it will be some time before the last is heard of the system.

In the health sciences those parts of the apothecaries' system that measure mass and volume are important. The basic unit of mass is the *grain* (gr), which in ancient times was the size of a grain of wheat taken from the center of a field. The basic unit of volume is the *minim* (m), which is roughly the size of a drop of water. But be careful: There are many different sized drops. Table 5.2 lists the apothecaries' system units of mass and volume, their relationships, and their symbols.

Table 5.2

Mass	Volume
60 grains (gr) = 1 dram ($\mathfrak{z}$)	60 minims = 1 fluid dram (f$\mathfrak{z}$)
8 drams = 1 ounce ($\mathfrak{z}$)*	8 fluid drams = 1 fluid ounce (f$\mathfrak{z}$)
	8 fluid ounces = 1 cup
	2 cups = 1 pint (pt)
	2 pints = 1 quart (qt)
	4 quarts = 1 gallon (gal)

*The apothecaries' ounce is 9.7% larger than the avoirdupois ounce.

The symbols for apothecaries' units are written to the left of the numerals. (Roman numerals may be used instead of Arabic numerals.) Therefore, the relationships in Table 5.2 would be translated as follows:

mass	*volume*	
gr 60 = $\mathfrak{z}$ 1	m 60 = f$\mathfrak{z}$ 1	At standard temperature and
$\mathfrak{z}$ 8 = $\mathfrak{z}$ 1	f$\mathfrak{z}$ 8 = f$\mathfrak{z}$ 1	pressure, 1 minim water has a
	f$\mathfrak{z}$ 8 = 1 cup	mass of 1 grain.
	2 cups = pt 1	
	pt 2 = qt 1	
	qt 4 = gal 1	

You must memorize these lists. These relationships generate many ratios that are necessary to convert one unit to another.

$$\text{gr } 60 \;=\; \text{ʒ } 1 \quad \text{is the ratio } \frac{\text{gr } 60}{\text{ʒ } 1}$$

$$\text{ʒ} 8 \;=\; \text{℥ } 1 \quad \text{is the ratio } \frac{\text{ʒ } 8}{\text{℥ } 1}$$

$$\text{♏} 60 \;=\; \text{fʒ } 1 \quad \text{is the ratio } \frac{\text{♏ } 60}{\text{fʒ } 1}$$

$$\text{fʒ } 8 \;=\; \text{f℥ } 1 \quad \text{is the ratio } \frac{\text{f℥ } 8}{\text{f℥ } 1}$$

Other ratios are easy to form:

$$\frac{\text{gr } 480}{\text{℥ } 1} \quad \text{is the ratio of grains to ounces.}$$

$$\frac{\text{♏ } 480}{\text{f℥ } 1} \quad \text{is the ratio of minims to fluid ounces.}$$

There is an important bridge between these units of mass and volume: *At standard temperature and pressure, 1 minim water has a mass of 1 grain.*

In medicine this means that water (or a water-based solution) is considered to have a ratio of

$$\text{♏ } 1_{H_2O} : \text{gr } 1$$

or

$$\frac{\text{♏ } 1}{\text{gr } 1}$$

This must be memorized.

Converting from One Unit to Another

To convert from one unit to another in the apothecaries' system, use the relationships in Table 5.2, which you have memorized, and choose the ratio that matches the conversion you are attempting to perform. (Before you begin the examples, review Sections 4.2 and 4.3.)

EXAMPLE 1

$$\mathfrak{z} \; \overline{\text{iss}} \; = \; \text{gr} \; \underline{\qquad}^*$$

To convert $1\frac{1}{2}$ drams to grains, you need a dram–grain ratio, which is $\mathfrak{z}\,1 : \text{gr}\,60$. Now set up the proportion and solve.

$$\frac{\mathfrak{z}\,1\frac{1}{2}}{\text{gr }x} = \frac{\mathfrak{z}\,1}{\text{gr }60}$$

$$1x = 1\frac{1}{2} \cdot 60$$

$$x = 90 \ \text{grains, or gr } 90$$

EXAMPLE 2

$$\mathfrak{z}\,\frac{1}{4} = \mathfrak{z} \; \underline{\qquad}$$

To convert $\frac{1}{4}$ dram to ounces, you need a dram–ounce ratio, which is $\mathfrak{z}\,8 : \mathfrak{z}\,1$. Now set up the proportion and solve.

$$\frac{\mathfrak{z}\,\frac{1}{4}}{\mathfrak{z}\,x} = \frac{\mathfrak{z}\,8}{\mathfrak{z}\,1}$$

$$8x = \frac{1}{4}$$

$$\frac{1}{8} \cdot 8x = \frac{1}{4} \cdot \frac{1}{8}$$

$$x = \frac{1}{32} \ \text{ounce, or } \mathfrak{z}\,\frac{1}{32}$$

EXAMPLE 3

$$\mathfrak{m}\,14 = \text{f}\mathfrak{z} \; \underline{\qquad}$$

To convert 14 minims to fluid drams, you need a minim–fluid dram ratio, which is $\mathfrak{m}\,60 : \text{f}\mathfrak{z}\,1$. Now set up the proportion and solve.

$$\frac{\mathfrak{m}\,14}{\text{f}\mathfrak{z}\,x} = \frac{\mathfrak{m}\,60}{\text{f}\mathfrak{z}\,1}$$

---

*Convert Roman numerals to Arabic for solutions and answers to problems.

EXAMPLE 3
continued

$$60x = 14$$

$$\frac{1}{60} \cdot 60x = 14 \cdot \frac{1}{60}$$

$$x = \frac{14}{60}$$

$$= \frac{7}{30} \text{ fluid dram, or } f\mathfrak{z}\,\frac{7}{30}$$

EXAMPLE 4

$$f\mathfrak{z}\,\overline{ss} = \mathfrak{m}\,\underline{\qquad}$$

To convert $\frac{1}{2}$ fluid ounce to minims, you need a fluid ounce–minim ratio, which is $f\mathfrak{z}\,1 : \mathfrak{m}\,480$. Now set up the proportion and solve.

$$\frac{f\mathfrak{z}\,\frac{1}{2}}{\mathfrak{m}\,x} = \frac{f\mathfrak{z}\,1}{\mathfrak{m}\,480}$$

$$1x = \frac{1}{2} \cdot 480$$

$$x = 240 \text{ minims, or } \mathfrak{m}\,240$$

EXAMPLE 5

$$\mathfrak{m}\,30_{H_2O} = gr\,\underline{\qquad}$$

To convert 30 minims water to grains, you need a minim–grain ratio, which is $\mathfrak{m}\,1 : gr\,1$. Now set up the proportion and solve.

$$\frac{\mathfrak{m}\,30}{gr\,x} = \frac{\mathfrak{m}\,1}{gr\,1}$$

$$x = 30 \text{ grains, or } gr\,30$$

Convert each unit as indicated. Show your work.

Exercises 5.2

1. gr xxx $= \mathfrak{z}\,\underline{\qquad}$ 2. gr 100 $= \mathfrak{z}\,\underline{\qquad}$
3. $\mathfrak{z}\,4 = \mathfrak{z}\,\underline{\qquad}$ 4. $\mathfrak{z}\,12 = \mathfrak{z}\,\underline{\qquad}$
5. $\mathfrak{m}\,xx = f\mathfrak{z}\,\underline{\qquad}$ 6. $\mathfrak{m}\,90 = f\mathfrak{z}\,\underline{\qquad}$
7. $f\mathfrak{z}\,6 = f\mathfrak{z}\,\underline{\qquad}$ 8. $f\mathfrak{z}\,24 = f\mathfrak{z}\,\underline{\qquad}$
9. gr 240 $= \mathfrak{z}\,\underline{\qquad}$ 10. gr 160 $= \mathfrak{z}\,\underline{\qquad}$
11. $f\mathfrak{z}\,20 = pt\,\underline{\qquad}$ 12. $f\mathfrak{z}\,8 = pt\,\underline{\qquad}$

13. ♏ 120 = f℥ _____   14. ♏ 360 = f℥ _____

15. ʒ s̄s̄ = gr _____   16. ʒ ii = gr _____

17. pt īs̄s̄ = f℥ _____   18. pt $\frac{1}{8}$ = f℥ _____

19. f℥ īs̄s̄ = ♏ _____   20. f℥ $\frac{1}{6}$ = ♏ _____

21. pt 16 = gal _____   22. gal $1\frac{1}{2}$ = pt _____

23. ℥ $\frac{1}{8}$ = ʒ _____   24. ℥ īs̄s̄ = ʒ _____

25. pt 5 = qt _____   26. pt 1 = qt _____

27. f℥ $\frac{1}{24}$ = fʒ _____   28. f℥ ii = fʒ _____

29. qt 2 = pt _____   30. qt $\frac{1}{4}$ = pt _____

31. ℥ $\frac{1}{8}$ = gr _____   32. ℥ s̄s̄ = gr _____

33. qt 6 = gal _____   34. gal $1\frac{1}{2}$ = qt _____

35. f℥ $\frac{1}{6}$ = ♏ _____   36. f℥ ii = ♏ _____

In problems 37–44 assume the liquid is water.

37. ♏ 25 = gr _____   38. gr 16 = ♏ _____
39. f℥ 100 = ℥ _____   40. ℥ 75 = f℥ _____
41. f℥ 4 = ℥ _____   42. ℥ 10 = f℥ _____
43. pt 1 = ℥ _____   44. qt 1 = ℥ _____

Section 5.3   **Medication Problems**

To solve medication problems involving the apothecaries' system of measurement, set up the problems in proportion form. Remember this basic formula:

Medication-on-hand ratio = desired-medication-ratio

The *medication on hand* is the drug as it is available in the supply cupboard or from the pharmacist. The *desired medication* is the amount of the drug that the doctor has ordered be given to the patient.

EXAMPLE 1

Desired medication: gr xx, PO; medication on hand: gr viiss tablets.

medication   desired
on hand   medication

$$\frac{\text{gr } 7\frac{1}{2}}{1 \text{ tablet}} = \frac{\text{gr } 20}{x \text{ tablets}}$$

$$7\frac{1}{2}x = 20$$

$$\frac{2}{15} \cdot \frac{15}{2} \cdot x = 20 \cdot \frac{2}{15}$$

$$x = \frac{8}{3}$$

$$= 2\frac{2}{3}$$

Give $2\frac{2}{3}$ tablets each gr $7\frac{1}{2}$

EXAMPLE 2

An oral liquid has gr $\frac{1}{4}$ drug in f℥ i. Administer gr $\frac{1}{8}$. Give f℥ _____ or f℈ _____.

    This is a two-part problem. First solve for fluid ounces and then convert to fluid drams.

desired   medication
medication   on hand

$$\frac{\text{gr } \frac{1}{8}}{\text{f℥ } x} = \frac{\text{gr } \frac{1}{4}}{\text{f℥ } 1}$$

$$\frac{1}{4}x = \frac{1}{8}$$

$$4 \cdot \frac{1}{4}x = \frac{1}{8} \cdot 4$$

$$x = \text{f℥ } \frac{1}{2}$$

EXAMPLE 2
continued

Now, change $f\!\!\!\!\text{ʒ}\,\frac{1}{2}$ to $f\!\!\!\!\text{ʒ}$.

$$\frac{f\!\!\!\!\text{ʒ}\,\dfrac{1}{2}}{f\!\!\!\!\text{ʒ}\,x} = \frac{f\!\!\!\!\text{ʒ}\,1}{f\!\!\!\!\text{ʒ}\,8}$$

$$1x = \frac{1}{2}\cdot 8$$

$$x = f\!\!\!\!\text{ʒ}\,4$$

Give $f\!\!\!\!\text{ʒ}\,\frac{1}{2}$ or $f\!\!\!\!\text{ʒ}\,4$ drug $(\text{gr}\,\frac{1}{4}/f\!\!\!\!\text{ʒ}\,1)$.

Exercises 5.3   Solve for the desired medication, being sure to label your answers. Show all your work.

1. Desired medication: gr xv; medication on hand: gr $2\frac{3}{4}$ tablets. Give _____ tablets.

2. Desired medication: gr iii; medication on hand: gr $\overline{\text{iss}}$ tablets. Give _____ tablets.

3. Give gr $\frac{1}{40}$ from tablets gr $\frac{1}{30}$. Give _____ tablets.

4. Give gr $\frac{1}{2}$ from tablets gr $\frac{1}{6}$. Give _____ tablets.

5. Give gr $\frac{1}{100}$ from tablets gr $\frac{1}{150}$. Give _____ tablets.

6. Give gr $\frac{1}{200}$ from tablets gr $\frac{1}{100}$. Give _____ tablets.

7. Tablets gr $\frac{1}{64}$ are available; administer gr $\frac{1}{24}$. Give _____ tablets.

8. Tablets gr $\frac{1}{3}$ are available; administer gr $\frac{1}{9}$. Give _____ tablets.

9. Tablets gr 5 are available; administer gr 12. Give _____ tablets.

10. Desired dose: gr $\frac{1}{400}$; medication on hand: gr $\frac{1}{100}$ tablets. Give _____ tablets.

11. An oral liquid has gr $\text{vii}\overline{\text{ss}}$ in $f\!\!\!\!\text{ʒ}$ ii; give gr x. How much will you administer?

12. An injectable liquid has gr $\frac{1}{150}$ drug in ♏ xv; give gr $\frac{1}{100}$. How much will you administer?

13. A vial on hand is labeled gr $\frac{1}{8}$/ ♏ xxx; give gr $\frac{1}{24}$. How much will you administer?

14. An injectable liquid has gr $\frac{1}{150}$ in ♏ xvi; give gr $\frac{1}{250}$. How much will you give?

15. Give gr $\frac{1}{8}$ from a solution labeled *gr* $\frac{1}{6}$: ♏ *xv*. How much will you give?

16. A syrup has gr $\frac{1}{6}$ drug in fʒ i; give gr $\frac{1}{8}$. Give fʒ _____ or ♏ _____ .

[5.1]

Review Exercises

1. You should remember the symbols for the first 30 Roman numerals. Especially note the different ways to write the first 10.

[5.2]
2. Memorize the apothecaries' symbols.

3. Memorize this chart:

mass	volume	
gr 60 = ʒ 1	♏ 60 = fʒ 1	At standard temperature and
ʒ 8 = ℥ 1	fʒ 8 = f℥ 1	pressure, 1 minim water has
	f℥ 8 = 1 cup	a mass of 1 grain.
	2 cups = pt 1	
	pt 2 = qt 1	
	qt 4 = gal 1	

[5.3]
Convert each unit as indicated:

4. gr xx = ʒ _____

5. ♏ 240 = f℥ _____

6. ℥ īss = ʒ _____

7. pt $\frac{1}{2}$ = f℥ _____

8. fʒ ii = ♏ _____

9. ♏ 15 of H_2O = gr _____

10. ℥ s͞s = gr _____

11. gal 1 = pt _____

Solve for desired medication. Be sure to label. Show work.

12. Desired medication: gr x; medication on hand: gr v tablets.

13. Desired medication: gr iii; medication on hand: gr s͞s in ♏ x.

14. An injectable liquid has gr $\frac{1}{100}$ in ♏ xvi; give gr $\frac{1}{150}$. How much will you give?

CHAPTER

6

The Metric System of Measurement (SI)

This chapter on the metric system of measurement introduces the International System of Units (SI), which is a universally accepted system of measurement. Understanding of and ability to use this system allow the student to understand measurements used in medical facilities in the United States and throughout the world.

Topics include:

Definitions, symbols, and values of the length, volume, and mass units of the SI system of measurement, and a method for easily changing from one unit to another

The interrelations of length, volume, and mass

Complicating factors of medication problems involving the metric system, including

> the unit of measurement in which the doctor orders medication differs from the unit on the available supply container

> more than one concentration of drug is available

Length, Volume, and Mass

The International System of Units, whose common abbreviation in all languages is SI, is the preferred system of measurement in science and medicine. In the United States reference to the metric system of measurement is reference to SI. SI uses seven base units (metre, kilogram, second, ampere, kelvin, mole, and candela), two supplementary units (radian and steradian), and a series of derived units for measuring all physical quantities. With this system one can express the measure of, or physical characteristics of, the longest distance, the smallest atomic particle, or the greatest mass in the universe.

Our primary concern is to learn the relationships that exist in measurement of length, volume, and mass. The reference unit of length is the *metre,* which is a generous yard (approximately 39.4 inches). The reference unit for volume is the *litre,* which is a generous quart (approximately 1.04 quart). The reference unit for mass is the *gram,* a measure slightly greater than 15 grains (Section 5.2).

The metric system is a base ten or decimal system (Section 2.2). Prefixes are used with the reference units to indicate this power of 10 relationship.* The prefixes, their positions, and the symbols for each from the kilo to the micro levels must be memorized.

Prefixes	Symbols	Prefixes	Symbols
kilo	(*k*)	centi	(*c*)
hecto	(*h*)	milli	(*m*)
deka	(*da*)	—	
metre (*m*), litre (*L*), gram (*g*)		—	
deci	(*d*)	micro	(*mc* or *μ*)

1. Because the metric system is a base ten system, decimal numerals rather than common fractions are used to express quantities.†

A System Table

---

*You may recognize some of these prefixes from the U.S. monetary system.
†The decimal system also facilitates the use of computers and calculators.

2. When speaking, say the full name of the unit, not the symbol. That is, say *kilometre* not *km*.

Table 6.1

Prefix	Symbol	Value	Length		Volume		Mass	
			Name	Symbol	Name	Symbol	Name	Symbol
mega	M	1 000 000 (10^6)	megametre	Mm	megalitre	ML	megagram	Mg
—	—	—	—		—		—	
*kilo	k	1 000 (10^3)	kilometre	km	kilolitre	kL	kilogram	kg
hecto	h	100 (10^2)	hectometre	hm	hectolitre	hL	hectogram	hg
deka	da	10	dekametre	dam	dekalitre	daL	dekagram	dag
		1	*metre*	m	*litre*	L	*gram*	g
*deci	d	0.1 (10^{-1})	decimetre	dm	decilitre	dL	decigram	dg
*centi	c	0.01 (10^{-2})	centimetre	cm	centilitre	cL	centigram	cg
*milli	m	0.001 (10^{-3})	millimetre	mm	millilitre	mL	milligram	mg
—	—	—	—		—		—	
—	—	—	—		—		—	
*micro	μ, mc	0.000 001 (10^{-6})	micrometre (micron)	μm	microlitre	μL	microgram	μg, mcg
—	—	—	—		—		—	
—	—	—	—		—		—	
nano	n	10^{-9}	nanometre		nanolitre		nanogram	
		10^{-10}	angstrom		—		—	

*The levels most frequently used in health-related occupations.

3. Refer to Table 6.1 and note that in any vertical column each unit has a value 10 times the value of the unit immediately below it in the column. Another way of saying the same thing is that it takes 10 of any unit to make 1 of the unit directly above it in the same column. For example,

> 10 millimetres = 1 centimetre
>
> 10 milligrams = 1 centigram
>
> 10 litres = 1 dekalitre
>
> 10 decigrams = 1 gram

This same power of 10 relationship exists in the decimal numeration system (Section 2.2).

4. Again referring to Table 6.1, there are unnamed levels in each column. Although they are unnamed, they also possess the

same power of 10 relationship just mentioned. When counting levels in any column, remember to count these unnamed levels as well.

To use Table 6.1 in any given column to change from a unit lower in the column (a smaller unit) to a unit higher in the column (a larger unit), follow these two rules:

1. Count the steps taken up the column.
2. Move the decimal point one position to the left for each step taken.

In each of the following examples, the change is from a unit lower in the column to a unit higher in the column; therefore, the decimal point moves to the left. Refer to the table to count steps.

$$1000 \text{ mm} = 100 \text{ cm}$$

EXAMPLE 1

One step up from mm to cm and the decimal point moves one position left.

$$1000 \text{ mm} = 10 \text{ dm}$$

EXAMPLE 2

Two steps up from mm to dm and the decimal point moves two positions left.

$$1000 \text{ mm} = 1 \text{ m}$$

EXAMPLE 3

Three steps up from mm to m and the decimal point moves three positions left.

$$1000 \text{ mm} = 0.001 \text{ km}$$

EXAMPLE 4

Six steps up from mm to km and the decimal point moves six positions left.

$$67 \text{ dL} = 6.7 \text{ L}$$

EXAMPLE 5

One step up from dL to L and the decimal point moves one position left.

EXAMPLE 6   250 cg = 2.5 g

Two steps up from cg to g and the decimal point moves two positions left.

EXAMPLE 7   250 μg = 0.25 mg

Three steps up from μg to mg and the decimal point moves three positions left.

> In any given column, to change from a unit higher in the column (a larger unit) to a unit lower in the column (a smaller unit), follow these two rules:
>
> 1. Count the steps taken down the column.
> 2. Move the decimal point one position to the right for each step taken.

In each of the following examples, the change is from a unit higher in the column to a unit lower in the column; therefore, the decimal point moves to the right. Refer to Table 6.1 to count steps.

EXAMPLE 8   2.5 mm = 2500 μm

Three steps down from mm to μm and the decimal point moves three positions right.

EXAMPLE 9   0.35 km = 3500 dm

Four steps down from km to dm and the decimal point moves four positions right.

EXAMPLE 10   0.06 L = 6 cL

Two steps down from L to cL and the decimal moves two positions right.

$$1 \text{ g} = 10 \text{ dg}$$

EXAMPLE 11

One step down from g to dg and the decimal point moves one position right.

$$1 \text{ g} = 100 \text{ cg}$$

EXAMPLE 12

Two steps down from g to cg and the decimal point moves two positions right.

$$1 \text{ g} = 1000 \text{ mg}$$

EXAMPLE 13

Three steps down from g to mg and the decimal point moves three positions right.

$$1 \text{ g} = 1\,000\,000 \text{ } \mu\text{g}$$

EXAMPLE 14

Six steps down from g to μg and the decimal point moves six positions right.

Again, it is essential that you memorize the prefixes from the kilo level to the micro level, not forgetting the reference units. You should be able to change from one unit to another in any column without referring to the table. Also, do not forget to count the unnamed levels.

There are more prefixes than those listed in Table 6.1. A complete list of them can be found in the Appendixes.

1. 1 L = _____ mL

2. 1 mL = _____ L   **Exercises 6.1**

3. 1 g = _____ mg

4. 1 mg = _____ g

5. 2.5 cg = _____ g

6. 2 kg = _____ g

7. 1.52 L = _____ mL

8. 0.8 g = _____ cg

9. 8 kg = _____ cg

10. 50 g = _____ kg

11. 2 m = _____ cm

12. 300 cm = _____ m

13. 25 cm = _____ mm

14. 25 cm = _____ dm

15. 25 cm = _____ m

16. 50 g = _____ dg

17. 1500 mg = _____ g

18. 25 000 mL = _____ L

19. 0.006 g = _____ mg

20. 20 mg = _____ μg

21. 20 mm = _____ μm   22. 0.065 g = _____ cg
23. 1625 mL = _____ L   24. 0.4 mg = _____ g
25. 5.4 L = _____ kL   26. 0.05 kg = _____ g
27. 1.5 L = _____ mL   28. 0.005 L = _____ mL
29. 0.0025 kg = _____ g   30. 10 g = _____ mg
31. 2 kg = _____ cg   32. 1000 mg = _____ g
33. 100 mL = _____ L   34. 1 m = _____ cm
35. 0.2 m = _____ mm   36. 2 mL = _____ dL
37. 0.5 L = _____ mL   38. 13 mm = _____ cm
39. 28 cm = _____ m   40. 265 cm = _____ m
41. 35.4 cm = _____ mm   42. 35.4 mm = _____ cm
43. 3 kg = _____ g   44. 65 mm = _____ dm
45. 65 mm = _____ cm   46. 150 mcg = _____ mg
47. 0.6 mg = _____ mcg   48. 45 mg = _____ g
49. 45 mg = _____ mcg   50. 0.0702 kg = _____ g
51. 0.9368 m = _____ mm   52. 37 mg = _____ g
53. 1 m = _____ km   54. 1 km = _____ m
55. 250 μg = _____ mg   56. 250 μg = _____ g
57. 50 mg = _____ g   58. 0.05 g = _____ mg
59. 0.05 g = _____ μg   60. 0.1 mg = _____ g

Section 6.2 ## The Interrelations of Length, Volume, and Mass

A cubic centimetre is a cube each edge of which is 1 centimetre long. The symbol for cubic centimetre is cm^3 (sometimes written *cc*).

Figure 6.1

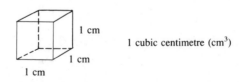

1 cubic centimetre (cm^3)

1 cubic centimetre is equivalent in volume to 1 millilitre.

$$1 \ cm^3 = 1 \ mL$$

Therefore, the names *cubic centimetre* and *millilitre* may be interchanged. For example, 1 litre equals 1000 millilitres, so 1 litre also equals 1000 cubic centimetres.

1 cubic centimetre (or 1 millilitre) of water at standard temperature and pressure has a mass equal to 1 gram.

$$1 \text{ cm}^3{}_{H_2O} = 1 \text{ mL}_{H_2O} = 1 \text{ g}$$

These relationships combined with those from Table 6.1 produce some interesting results. For example, if

$$1 \text{ cm}^3{}_{H_2O} = 1 \text{ mL}_{H_2O} = 1 \text{ g}$$

then

$$1000 \text{ cm}^3{}_{H_2O} = 1000 \text{ mL}_{H_2O} = 1000 \text{ g}$$

or

$$1 \text{ dm}^3{}_{H_2O} = 1 \text{ L}_{H_2O} = 1 \text{ kg}$$

We know that 1000 millilitres equals 1 litre and that 1000 grams equals 1 kilogram. Now, consider 1000 cubic centimetres. If 1000 cubic centimetres are arranged in a large cube, then each edge of the cube would be 10 centimetres long. And since 10 centimetres equals 1 decimetre, each edge of the cube would be 1 decimetre long and the large cube itself would be a cubic decimetre (dm^3). Or

$$1000 \text{ cm}^3 = 1 \text{ dm}^3$$

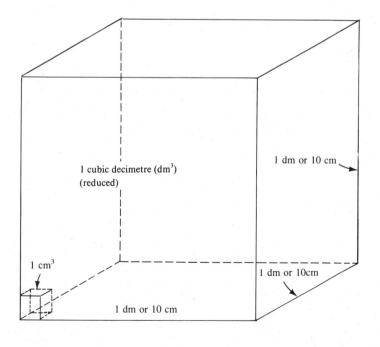

Figure 6.2

1 cubic decimetre (dm^3)
(reduced)

1 dm or 10 cm

1 cm^3

1 dm or 10cm

1 dm or 10 cm

We know that

$$1000 \text{ cm}^3 = 1 \text{ dm}^3$$

and we know that

$$1000 \text{ mL} = 1 \text{ L}$$

therefore

$$1 \text{ dm}^3 = 1 \text{ L}$$

Also, we know that

$$1000 \text{ mL}_{H_2O} = 1000 \text{ g}$$

and that

$$1000 \text{ g} = 1 \text{ kg}$$

therefore

$$1 \text{ dm}^3{}_{H_2O} = 1 \text{ L}_{H_2O} = 1 \text{ kg}$$

We have shown that 1 litre water at standard temperature and pressure has a mass of 1 kilogram and is equivalent in volume to 1 cubic decimetre.

In summation if we have water at standard temperature and pressure, then

$$\begin{array}{cc} volume & mass \\ 1 \text{ L}_{H_2O} = & 1 \text{ kg} \\ 1 \text{ mL (cm}^3)_{H_2O} = & 1 \text{ g} \\ 1 \text{ } \mu\text{L}_{H_2O} = & 1 \text{ mg} \end{array}$$

Frequently in health-related occupations these relationships are used for water-based solutions as well as for water. So 2.5 L dilute sodium chloride solution has an approximate mass of 2.5 kilograms. And if orders ask for 5 grams water-based solution, in most cases you may administer 5 mL or 5 cm^3.

Dosages frequently are determined per square metre body surface. A square metre (m^2) is a square each edge of which is 1 metre long. *Body surface* is the total amount of skin covering a body. For example, an adult who is 65 inches tall and whose mass is 130 pounds has a body surface of about 1.65 square metres. A child whose height is 41 inches and whose mass is 37 pounds has a body surface of 0.69 square metres. To determine the body surface area of a patient, see the West Nomogram in Appendix.

1. $2L = $ _____ dm^3
2. $2L = $ _____ cm^3
3. $2L = $ _____ mL
4. $2L = $ _____ cc
5. $2L_{H_2O} = $ _____ kg
   $= $ _____ g
6. $0.06\ cm^3_{H_2O} = $ _____ g
   $= $ _____ mg
7. $1.05\ mL_{H_2O} = $ _____ g
   $= $ _____ mg
8. $25\ mL = $ _____ cm^3
9. $1\ mL = $ _____ L
10. $1\ cm^3 = $ _____ L
11. $0.003\ L = $ _____ cm^3
    $= $ _____ mL
12. $0.003\ L = $ _____ dm^3
13. $0.003\ dm^3_{H_2O} = $ _____ kg
    $= $ _____ g
    $= $ _____ mg
14. $15\ \mu L_{H_2O} = $ _____ mg

Medication Problems

Many medication problems using the metric system were solved in Section 4.3, which should be reviewed at this time. Here we work with the same type of problems, though a bit more complicated.

Medication frequently is ordered for a patient in units that differ from those on the label of the supply container. In this case the problem is twofold. *You must first change the desired medication units to the units on the label of the supply container.* Second, you must solve the medication problem.

When the Units the Doctor Orders Differ from the Units on the Supply Container

EXAMPLE 1

Order: 1.2 g daily as 4 separate doses; medication on hand: 100 mg tablets.

This is a two-step problem. First, change the desired medication units (1.2 g) to the units on the label of the supply container (mg).

$$1.2\ g = 1200\ mg\ \text{total daily in 4 divided doses}$$

Second, solve the medication problem

$$\underset{\text{medication on hand}}{\frac{100\ mg}{1\ tab}} = \underset{\text{desired medication}}{\frac{1200\ mg}{x\ tabs}}$$

EXAMPLE 1
continued

$$100x = 1200$$

$$x = 12 \text{ tablets (100 mg) daily in}$$
4 divided doses

$$12 \div 4 = 3$$

Administer 3 tabs (100 mg), QID (see Appendix).

EXAMPLE 2

Order: indocyanine green 500 mcg/kg body mass; mass of patient is 60 kg. Medication on hand: indocyanine green 5 mg/mL. Determine the desired medication (Section 4.3, Example 6):

$$\frac{500 \text{ mcg}}{1 \text{ kg}} = \frac{x \text{ mcg}}{60 \text{ kg}}$$

$$x = 30\ 000 \text{ mcg indocyanine green for a}$$
60-kg patient

Change the desired medication units to available units:

$$30\ 000 \text{ mcg} = 30 \text{ mg indocyanine green for a}$$
60-kg patient

Solve the medication problem:

$$\frac{\overset{medication\ on\ hand}{\text{indocyanine green 5 mg}}}{1 \text{ mL}} = \frac{\overset{desired\ medication}{\text{30 mg indocyanine green}}}{x \text{ mL}}$$

$$5x = 30$$

$$x = 6 \text{ mL}$$

Administer 6 mL indocyanine green (5 mg/mL) to a 60-kg patient.

When more than one concentration of drug is available. A second complicating factor arises when there is more than one container of the desired drug on the supply shelf and each is of a different concentration. Keeping in mind the comfort and welfare of the patient, you must make a choice of concentration.

EXAMPLE 3

Desired medication: ferrous sulfate 1.2 g total daily as 3 separate doses; medication on hand: ferrous sulfate tablets 200 mg, 300 mg, and 325 mg. Change the desired medication units to the units in the supply container:

$$\text{ferrous sulfate } 1.2 \text{ g} = 1200 \text{ mg total daily}$$
$$\text{as 3 separate doses or 400 mg}$$
$$\text{per dose}$$

Choose the drug concentration:

200 mg ferrous sulfate tablets

Solve the medication problem:

$$\frac{\underset{\text{ferrous sulfate 200 mg}}{\text{medication on hand}}}{1 \text{ tab}} = \frac{\underset{\text{400 mg ferrous sulfate}}{\text{desired medication}}}{x \text{ tab}}$$

$$x = 2 \text{ tabs (200 mg)}$$

Administer 2 tabs ferrous sulfate (200 mg), TID.

Solve for the information requested, being sure to label and check your answers.

1. Desired medication: 4 g total daily as 4 divided doses; medication on hand: 500 mg tablets. How many tablets will you give per dose?

2. Desired medication: 1.2 g; medication on hand: 300 mg tablets. How many tablets will you administer?

3. Desired medication: 1.5 g; medication on hand: 500 mg tablets. How many tablets will you give?

4. Desired medication: 80 mg; medication on hand: 0.16 g tablets. How many tablets will you administer?

5. Desired medication: 1000 mg; medication on hand: 2 g tablets. How many tablets will you give?

6. Desired medication: 300 mcg; medication on hand: 0.4 mg tablets. How many tablets will you give?

7. Desired medication: 150 mcg; medication on hand: 0.3 mg tablets. How many tablets will you administer?

8. Desired medication: 100 mg; medication on hand: 0.05 g per mL. How much will you administer?

9. Desired medication: 100 mg, PO; medication on hand: 0.2 g per 5 mL. How much will you administer?

10. Desired medication: 0.02 mg; medication on hand: 15 mcg per mL. How much will you administer?

11. Desired medication: 2 g; medication on hand: 650 mg/5 mL. How much will you administer?

12. Desired medication: 3.2 g, PO, total daily as 4 divided doses; medication on hand: 400 mg/5 mL suspension. How much will you give?

13. You've been asked to administer a total daily dose of 300 mg of simethicone after meals and before bedtime. Simethicone tablets 40 mg, 50 mg, and 80 mg scored are available. How many of which strength tablet should you administer per dose?

14. For rheumatoid arthritis 3.2 g aspirin daily in 4 divided doses is recommended. Aspirin tablets 300 mg, 325 mg, 400 mg, and 454 mg are available. How many of which strength tablet will you give/dose?

15. Reserpine 0.25 mg total daily in 2 divided doses is ordered. Reserpine tablets 100 mcg, 125 mcg, 200 mcg, and 250 mcg are available. How many of which strength tablet will you administer per dose?

16. Sodium bicarbonate 0.9 g is ordered. Sodium bicarbonate tablets 300 mg and 600 mg are available. Which and how many will you use?

17. Methyclothiazide 150 mcg/kg of body mass is ordered. The patient has a mass of 25 kg. Methyclothiazide tablets 2.5 mg scored and 5 mg scored are available. Which and how many will you use?

18. Gantrisin syrup 2 g/m^2 body surface is ordered. The patient has a body surface of 0.45 m^2. Gantrisin syrup is available 500 mg/5 mL. How much will you administer?

19. You've been asked to administer hydrocodone bitartrate 20 mg/m^2 body surface, PO, daily in 3 divided doses. The body surface of the patient is 0.76 m^2. Hydrocodone bitartrate syrup is available in 1.66 mg/5 mL, 2.5 mg/5 mL, and

5 mg/5 mL strengths. How much of which strength will you administer/dose?

20. A total initial dose of sulfadiazine and sulfamerazine 2 g/m^2 body surface has been requested. The patient has 0.6 m^2 body surface. Sulfadiazine and sulfamerazine are available as an oral suspension 167 mg of each per 5 mL. How much will you administer?

[6.1]   **Review Exercises**

Memorize the SI symbols, the prefix names, and the prefix positions from kilo to micro levels.

kilo (k)

hecto (h)

deka (da)

  metre (m), litre (L), gram (g)

deci (d)

centi (c)   To change from one unit to another:

milli (m)   1. the decimal point moves left one position
                       for each step up a column;
_____
                   2. the decimal point moves right one position
_____   for each step down a column.

micro (mc, μ)

1. 2 g = _____ mg   2. 50 μg = _____ mg

3. 1.5 L = _____ mL   4. 0.03 g = _____ μg

5. 1.05 cm = _____ mm   6. 1.5 cm = _____ m

7. 22 g = _____ kg   8. 0.5 mL = _____ dL

9. 12 μm = _____ mm   10. 0.5 mL = _____ μL

[6.2]

11. 5 mL$_{H_2O}$ = _____ g   12. 5 μL$_{H_2O}$ = _____ mg

13. 5 mL = _____ cm^3 14. 5 L$_{H_2O}$ = _____ kg

[6.3]

Calculate dosages:

15. Desired medication: 1.2 g total daily as 3 equally divided doses. Medication on hand: tablets 300 mg, 400 mg, and 500 mg.

16. Order: Atropine sulfate 10 μg/kg, IM, q6h; patient weight is 21 kg.

    Available: Atropine sulfate for injection: 0.05 mg/mL, 0.1 mg/mL, and 0.3 mg/mL.

17. Order: promethazine 10 mg/m^2, PO, q4h; patient body surface area is 0.95 m^2.

    Available: promethazine 6.25 mg/5 mL and 25 mg/5 mL.

Equivalents

This chapter contains discussion of equivalencies among measurement systems: the ancient, awkwardly organized apothecaries' system; the recent, precisely defined metric (SI) system; household units; and U.S. Customary units. Medication may be prescribed in apothecaries' units and administered from a supply container labeled with metric units. Equipment may be labeled with metric units (some not SI standard symbols), apothecaries' units, household units, or with U.S. Customary units. Medication may be administered in the home using home equipment, and the health provider must be the interpreter for the layperson who is to administer the drug. Emphasis is placed on use of universally available equipment as an aide in remembering equivalencies.

Topics include:

Metric, apothecaries', and household equivalencies

Metric and U.S. Customary equivalencies

Medication problems involving these equivalencies and their complications, including

      total medication must be calculated

      best drug concentration decisions must be made

      extraneous information is included

Section 7.1   **Metric–Apothecaries–Household Equivalents**

The solution of health-related problems will be more accurate when performed only in the metric (SI) system of measurement because the metric–apothecaries'–household equivalents are at best approximate. However, it probably will be years before the dual and triple systems of measurement are phased out.

There are dozens of approximate ratios metric to apothecaries' and vice versa. You will find certain of these to be more important than others. Measuring instruments commonly found in all places where medicine is practiced are excellent sources of equivalent ratios involving metric, apothecaries', and household measure. Consider the 1 fluid ounce medicine cup (Figure 7.1), which is equal in volume to the 1 fluid ounce found on kitchen measuring cups.

Figure 7.1

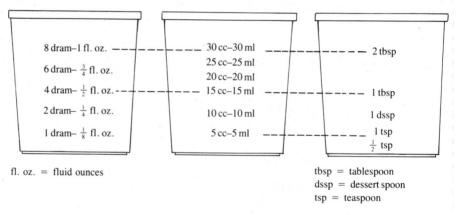

8 dram–1 fl. oz.   30 cc–30 ml   2 tbsp

25 cc–25 ml

6 dram– $\frac{3}{4}$ fl. oz.   20 cc–20 ml

4 dram– $\frac{1}{2}$ fl. oz.   15 cc–15 ml   1 tbsp

2 dram– $\frac{1}{4}$ fl. oz.   10 cc–10 ml   1 dssp

1 dram– $\frac{1}{8}$ fl. oz.   5 cc–5 ml   1 tsp

$\frac{1}{2}$ tsp

fl. oz. = fluid ounces

tbsp = tablespoon
dssp = dessert spoon
tsp = teaspoon

There are many ratios on the cup, some of which you already know. Look at the top level of the cup and see that

$$f\!\!\!3\; 1 = f\!\!\!3\; 8 = 30 \text{ mL} = 30 \text{ cm}^3 = \quad 2 \text{ tbsp}$$

(apothecaries')   (metric)   (household)

These approximate equivalencies become important bridges when converting from one system to another. Some ratios from these equivalencies are

$$\frac{1 \text{ mL}}{1 \text{ cm}^3} \qquad \frac{f\!\!\!3\; 1}{30 \text{ mL}} \qquad \frac{f\!\!\!3\; 1}{f\!\!\!3\; 8} \qquad \frac{30 \text{ mL}}{2 \text{ tbsp}} \qquad \frac{f\!\!\!3\; 1}{2 \text{ tbsp}}$$

It is easy to find many others on the medicine cup, but two of these and one more ratio to be found farther down the cup are key ratios and should be memorized:

Figure 7.2

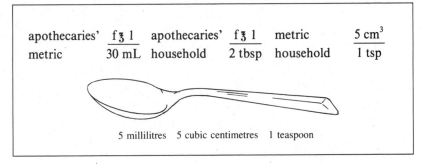

apothecaries' $\dfrac{\text{f} \, \text{ʒ} \, 1}{30 \text{ mL}}$ apothecaries' $\dfrac{\text{f} \, \text{ʒ} \, 1}{2 \text{ tbsp}}$ metric $\dfrac{5 \text{ cm}^3}{1 \text{ tsp}}$
metric   household   household

5 millilitres   5 cubic centimetres   1 teaspoon

Look again at the third ratio—(5 cubic centimetres: 1 teaspoon). See Section 6.3 exercises 9, 11, 12, 18, 19, and 20. Note the number of drugs that are to be administered orally (syrup, suspension, elixir) and that have concentrations expressed per 5 millilitres. The concentrations are expressed per teaspoon for easy administration at home.

These are not the only equivalency ratios needed. For example, any apothecaries' unit equivalent to 1 fluid ounce can be substituted for 1 fluid ounce. Using this and the fluid ounce–millilitre ratio (1 fluid ounce: 30 millilitres), you can see that

$$\frac{\text{f} \, \text{ʒ} \, 1}{30 \text{ mL}} = \frac{\text{f} \, \text{ʒ} \, 8}{30 \text{ mL}} = \frac{\text{ʍ} \, 480}{30 \text{ mL}}$$

which reduced becomes

$$\frac{\text{ʍ} \, 16^*}{1 \text{ mL}} \quad \text{or} \quad \frac{\text{ʍ} \, 1}{0.06 \text{ mL}}$$

The last two ratios are very important. As you recall, a minim is roughly the size of a drop, and we now can say *16 (or 15) minim-sized drops equal 1 millilitre or 1 cubic centimetre.* This ratio appears on another common instrument, the 1 mL or 1 cm³ (cc) syringe, which sometimes is called a *tuberculin syringe* (Figure 7.3).

---

*You may find ʍ15:1 mL used instead of ʍ16:1 mL.

Figure 7.3

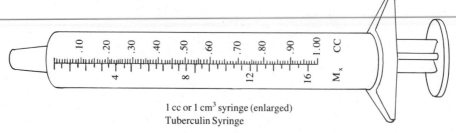

1 cc or 1 cm³ syringe (enlarged)
Tuberculin Syringe

Also needed is a ratio for mass, and perhaps the most frequently used is

$$\frac{\text{gr } 1}{60\text{–}65 \text{ mg}}$$

1 grain is approximately equivalent to any measurement in the 60 to 65 milligram range; gr 1 : 65 mg is most accurate, however. From this one ratio you can generate others as needed. For example,

$$\frac{\text{gr } 1}{60 \text{ mg}} = \frac{\text{gr } 1}{60\,000 \text{ mcg}} = \frac{\text{gr } 1}{0.06 \text{ g}}$$

Notice how much the last ratio is like the minim–millilitre ratio:

$$\frac{\text{gr } 1}{0.06 \text{ g}} \quad \frac{\text{m } 1}{0.06 \text{ mL}}$$

Memorize both the grain–milligram and minim–millilitre ratios:

$$\frac{\text{m } 16 \text{ (or m } 15)}{1 \text{ mL}} \quad \text{or} \quad \frac{\text{m } 1}{0.06 \text{ mL}} \qquad \frac{\text{gr } 1}{60\text{–}65 \text{ mg}} \quad \text{or} \quad \frac{\text{gr } 1}{0.06 \text{ g}}$$

There are a total of five ratios that must be memorized from this section. Remember that most of these can be found on the medicine cup and 1-mL syringe. These ratios, or ones generated from them, are used to solve almost all conversion problems involving metric, apothecaries', or household units.

EXAMPLE 1 The pancreas excretes about 32 fluid ounces of digestive juices a day. How many millilitres does the pancreas excrete?

EXAMPLE 1
continued

$$\frac{f\!\!\!Ʒ\,1}{30 \text{ mL}} = \frac{f\!\!\!Ʒ\,32}{x \text{ mL}}$$

$$x = 30 \cdot 32$$

$$x = 960 \text{ mL excreted per day}$$
$$\text{(about 1 litre)}$$

EXAMPLE 2

$$♏\,80 = \underline{\hspace{1cm}} \text{ mL}$$

$$\frac{♏\,16}{1 \text{ mL}} = \frac{♏\,80}{x \text{ mL}}$$

$$16x = 80$$

$$x = 5.0 \text{ mL}$$

Always express metric answers in decimal form; apothecaries and household answers are expressed in common fraction form.

EXAMPLE 3

$$Ʒ\,\text{ii} = \underline{\hspace{1cm}} \text{ tbsp}$$

$$\frac{Ʒ\,8}{2 \text{ tbsp}} = \frac{Ʒ\,2}{x \text{ tbsp}}$$

$$8x = 4$$

$$x = \frac{1}{2} \text{ tbsp}$$

EXAMPLE 4

$$\text{gr}\,\frac{1}{8} = \underline{\hspace{1cm}} \text{ mg}$$

$$\frac{\text{gr}\,\frac{1}{8}}{x \text{ mg}} = \frac{\text{gr}\,1}{64 \text{ mg*}}$$

$$x = 8 \text{ mg}$$

---

*This is in the acceptable 60 to 65 milligram range.

EXAMPLE 5

$$12 \text{ mL} = \underline{\hspace{1cm}} \text{ tsp}$$

$$\frac{12 \text{ mL}}{x \text{ tsp}} = \frac{5 \text{ mL}}{1 \text{ tsp}}$$

$$5x = 12$$

$$x = 2\frac{2}{5}, \text{ or about } 2\frac{1}{2} \text{ tsp}$$

EXAMPLE 6

The pancreas has a mass of about 3 ounces. This is how many grams?

$$\text{ʒ } 3 = \underline{\hspace{1cm}} \text{ g}$$

$$\frac{\text{ʒ } 1}{30 \text{ g}} = \frac{\text{ʒ } 3}{x \text{ g}}$$

$$x = 90 \text{ grams}$$

Exercises 7.1   Solve for the requested information. Be sure to show your work and check your answers.

1. $0.25 \text{ g} = \text{gr} \underline{\hspace{1cm}}$

2. $0.04 \text{ g} = \text{gr} \underline{\hspace{1cm}}$

3. $\text{gr vi} = \underline{\hspace{1cm}} \text{ g}$

4. $\text{gr iii}\overline{\text{ss}} = \underline{\hspace{1cm}} \text{ g}$

5. $0.75 \text{ g} = \text{gr} \underline{\hspace{1cm}}$

6. $\text{gr } \dfrac{3}{4} = \underline{\hspace{1cm}} \text{ mg}$

7. $\text{gr ii} = \underline{\hspace{1cm}} \text{ mg}$

8. $750 \text{ mg} = \text{gr} \underline{\hspace{1cm}}$

9. $0.1 \text{ mg} = \text{gr} \underline{\hspace{1cm}}$

10. $\text{gr} \dfrac{1}{200} = \underline{\hspace{1cm}} \text{ mcg}$

11. $\text{gr} \dfrac{1}{300} = \underline{\hspace{1cm}} \text{ mcg}$

12. $600 \text{ mcg} = \text{gr} \underline{\hspace{1cm}}$

13. $\text{gr} \dfrac{1}{12} = \underline{\hspace{1cm}} \text{ g}$

14. $1.25 \text{ mL} = \text{m} \underline{\hspace{1cm}}$

15. $\text{fʒ } \dfrac{1}{3} = \text{m} \underline{\hspace{1cm}}$

16. $0.24 \text{ mL} = \text{m} \underline{\hspace{1cm}}$

17. $\text{m } 80 = \underline{\hspace{1cm}} \text{ tsp}$

18. $\text{fʒ viii} = \underline{\hspace{1cm}} \text{ tbsp}$

19. $\text{fʒ i}\overline{\text{ss}} = \underline{\hspace{1cm}} \text{ tsp}$

20. $3 \text{ tbsp} = \underline{\hspace{1cm}} \text{ mL}$

21. $1 \text{ tbsp} = \text{m} \underline{\hspace{1cm}}$

22. $150 \text{ mL} = \underline{\hspace{1cm}} \text{ cm}^3$

23. $150 \text{ mcg} = \text{gr} \underline{\hspace{1cm}}$

24. $\text{fʒ viii} = \underline{\hspace{1cm}} \text{ tsp}$

25. $50 \text{ mL} = \text{fʒ} \underline{\hspace{1cm}}$

26. $50 \text{ mL} = \text{fʒ} \underline{\hspace{1cm}}$

27. 50 mL = _____ tbsp

28. gr $\overline{ss}$ = _____ mg

29. gr $\dfrac{1}{60}$ = _____ mg

30. 450 μg = gr _____

Metric-U.S. Customary Equivalents   Section 7.2

There are metric–U.S. customary approximate equivalents that are very important in the health sciences. You should memorize these two:

$$\dfrac{1 \text{ kg}}{2.2 \text{ lbs}} \text{ (avoirdupois)} \qquad \text{(for converting larger masses)}$$

$$\dfrac{1 \text{ inch}}{2.54 \text{ cm}} \qquad \text{(for converting linear measures)}$$

The second ratio generates other ratios for use in measuring smaller lengths:

$$\dfrac{1 \text{ in}}{2.54 \text{ cm}} = \dfrac{1 \text{ in}}{25.4 \text{ mm}} = \dfrac{1 \text{ in}}{25\ 400 \ \mu\text{m (microns)}}$$

A patient has a mass of 115 lbs. How many kilograms is this?   **EXAMPLE 1**

$$\dfrac{1 \text{ kg}}{2.2 \text{ lbs}} = \dfrac{x \text{ kg}}{115 \text{ lbs}}$$

$$2.2x = 115$$

$$x = 52.3 \text{ kg}$$

The pancreas is about 6 in long. How many centimetres is this?   **EXAMPLE 2**

$$\dfrac{1 \text{ in}}{2.54 \text{ cm}} = \dfrac{6 \text{ in}}{x \text{ cm}}$$

$$x = 15.24, \text{ or } 15.2 \text{ cm long}$$

Red blood corpuscles (erythrocytes) have an average diameter of 7.7 μm. What part of an inch is this?   **EXAMPLE 3**

EXAMPLE 3
continued

$$\frac{1\ in}{25\ 400\ \mu m} = \frac{x\ in}{7.7\ \mu m}$$

$$25\ 400x = 7.7$$

$$x = 0.0003\ in$$

EXAMPLE 4

This is a two-part problem. A syrup, 2 molar, is ordered 2.5 mL/kg body mass/24 hr in 3 divided doses. How much will be administered a 152-lb patient? First change 152 lbs to kilograms:

$$\frac{1\ kg}{2.2\ lbs} = \frac{x\ kg}{152\ lbs}$$

$$x = 69\ kg$$

Now solve for medication:

$$\frac{2.5\ mL}{1\ kg} = \frac{x\ mL}{69\ kg}$$

$$x = 172.5\ mL\ syrup\ total\ daily$$

$$172.5 \div 3 = 57.5\ mL\ syrup\ per\ dose$$

Exercises 7.2

Solve for the requested information. Show your work and label and check the answers.

1. 80 lbs = _____ kg   2. 145 lbs = _____ kg

3. 8.1 kg = _____ lbs   4. 53 kg = _____ lbs

5. 12 in = _____ cm   6. 70 in = _____ cm

7. 34 in = _____ cm   8. 100 cm = _____ in

9. 16.8 cm = _____ in   10. 1.6 in = _____ μm

11. $\frac{1}{8}$ in = _____ μm 12. 45 microns = _____ in

13. Acetylsalicylic acid 65 mg/kg body mass/24 hr as 4 divided doses is ordered. How much will you administer to a 55-lb patient in 24 hours? (See Section 4.3, Example 6.) What will each dose be?

14. The doctor has ordered meperidine hydrochloride 1.6 mg/kg body mass/24 hr as 6 divided doses. How much will you administer to a 42-lb patient in 24 hours? What will each dose be?

15. Digoxin is ordered administered intravenously, 12 mcg/kg body mass. How much will a 92-lb patient receive?

16. The doctor has ordered epinephrine 1 : 1000 solution 0.01 mL/kg body mass. How much will a 63-lb patient receive?

17. Magnesium sulfate 50% solution is ordered 0.2 mL/kg body mass. How much will a 188-lb patient receive?

18. The doctor has ordered glycerin 50% solution 1.2 mL/kg body mass. How much will a 122-lb patient receive?

19. The average diameter of capillaries is about 8 μm. What part of an inch is this?

20. The aorta is 3 cm in diameter at its commencement. How many inches is this?

21. A medium-sized artery has a diameter of 3.4 mm. This is what part of an inch?

22. The capacity of the right atrium of the heart is about 57 mL. This is how many fluid ounces?

23. The average length of a female small intestine is 23 ft, 4 in. How many metres is this?

24. The large intestine is about 1.5 m long. How many inches is this?

25. A moderately full bladder holds 0.5 L. This is how many fluid ounces?

26. The male heart may be as large as 340 g. This is how many pounds?

27. The female liver is about 1.3 kg. This is how many pounds?

28. The spermatozoön has a head measurement of about 5 μm. This is what part of an inch?

29. The human ovum is about 0.14 mm in diameter. This is what part of an inch?

Medication Problems

Section 7.3

The medication exercises in this section are similar to those in earlier sections with one exception. These problems combine the metric, apothecaries', household, and U.S. Customary systems of measurement.

References to the prior sections are given with the problem sets. The following may be helpful as well:

1. If the medication is ordered in units that differ from the units on the label of the supply container, change the desired medication units to available units (Section 6.3).

2. To compare quantities that are expressed in different units, first change the quantities to the same unit. It will not matter which quantity is changed.

EXAMPLE 1

250 mcg is what part of gr $\frac{1}{120}$? To begin, change gr $\frac{1}{120}$ to mcg:

$$\frac{\text{gr}\,\dfrac{1}{120}}{x \text{ mcg}} = \frac{\text{gr } 1}{60\ 000 \text{ mcg}}$$

$$x = 500 \text{ mcg}$$

Our problem now is *250 mcg is what part of 500 mcg?*

$$\frac{250 \text{ mcg}}{500 \text{ mcg}} = \frac{1}{2} \qquad\qquad \text{(Section 1.7)}$$

250 mcg is one half gr$\frac{1}{120}$.

3. There is extra or unneeded information in some of the problems, just as there is unused information on a supply bottle.

Exercises 7.3   For Exercises 1–6 review Sections 1.7 and 7.1. Be sure to use the same units for comparison.

1. gr $\frac{1}{150}$ is how many times 200 mcg?
2. 0.4 g is how many times gr iii?
3. 300 mcg is what part of gr $\frac{1}{100}$?
4. gr $1\frac{1}{2}$ is what part of 0.3 g?
5. gr 10 is how many times 60 mg?
6. 0.15 mg is what part of gr $\frac{1}{200}$?

For Exercises 7–10 review Sections 6.3 and 7.1.

7. Tablets 0.2 g and 0.3 g are available. Which will you select to administer gr iii?

8. Phenobarbital tablets 32 mg, 65 mg, and 0.1 g are available. Which will you select to administer
    a.   gr $\overline{ss}$?   b.   gr $\overline{iss}$?

9. Suppositories 8 mg, 15 mg, 30 mg, 60 mg, 0.1 g, and 0.12 g are available. Which will you select to administer

    a.   gr $\frac{1}{8}$?   b.   gr ii?

10. Drug tablets 2.5 mg, 5 mg, and 10 mg are available. Which will you choose to administer

    a.   gr $\frac{1}{90}$?   b.   gr $\frac{1}{12}$?

For Exercises 11–16 review Sections 6.3, 7.1, and 7.2.

11. Caffeine and sodium benzoate is available for injection 250 mg per mL in 2-mL ampuls. Administer gr viii.

12. You have been asked to administer morphine sulfate gr $\frac{1}{6}$ subcutaneously to a pediatric patient. A single dose should not exceed 15 mg. Morphine sulfate 15 mg per mL is available. How much will you administer?

13. A 60-lb child is to receive, over a period of 24 hours, a dose of isotonic sodium chloride equivalent to 3% of body mass. How many millilitres will you administer? (See Section 6.2.) Consider 1 mL isotonic sodium chloride to have a mass of approximately 1 g.

14. Using home equipment, administer 5 mg hydrocodone bitartrate. Hydrocodone bitartrate 2.5 mg per mL is available.

15. The doctor has ordered 2 tsp elixir to be administered four times daily, 15 minutes before meals and at bedtime. How long will a 375 mL bottle of elixir last?

16. Using home equipment, how will you measure
    a.   f$\mathfrak{z}$ i?   b.   12 mL?
    c.   10 g water-based solution?   d.   5 mL?

Sections 1.9 and 2.2 will help with Exercises 17 and 18.

17. Which is larger, gr $\frac{1}{120}$ or 400 mcg?

18. Which is larger, 1.5 g or gr $\overline{iss}$?

Review Exercises   [7.1]
Memorize:

$$\frac{f\!\!\!\!\!\text{ʒ}\,1}{30\ \text{mL}}, \quad \frac{f\!\!\!\!\!\text{ʒ}\,1}{2\ \text{tbsp}}, \quad \frac{5\ \text{mL}}{1\ \text{tsp}}, \quad \frac{\text{gr}\ 1}{60\text{--}65\ \text{mg}}, \quad \frac{\text{m}\ 16}{1\ \text{mL}}$$

1. $0.02\ \text{g} = \text{gr}$ _____

2. $\text{gr ii} =$ _____ mg

3. $15\ \text{mL} =$ _____ tsp

4. $\text{m}\ 12 =$ _____ mL

5. $f\!\!\!\!\!\text{ʒ}\ 1\frac{1}{2} =$ _____ tbsp

6. $\text{gr}\ \frac{1}{2} =$ _____ mcg

[7.2]

7. $90\ \text{lbs} =$ _____ kg

8. $18\ \text{in} =$ _____ cm

9. $1.5\ \text{in} =$ _____ microns

10. $15\ \text{g}_{H_2O} =$ _____ mL
        $=$ _____ tsp

Calculate dosages:

11. Order: Theophylline 10 mg/kg daily as 2 equally divided doses; patient weight 55 lbs. Available: Theophylline suppositories 125 mg, 250 mg, and 500 mg.

12. Order: Atropine sulfate 10 mcg/kg, PO, q6h; patient weight 71 lbs. Available: Atropine sulfate tablets 0.3 mg, 0.4 mg, and 0.6 mg.

[7.3]

13. $\text{gr}\ \frac{1}{8}$ is what part of 24 mg?

14. $\text{gr}\ \frac{1}{100}$ is how many times 300 mcg?

15. Express in grams 2% of 90 lbs.

16. Calculate dosage: Order: promethazine 0.5 mg/kg, PO, at bedtime; patient weight 52 lbs. Available: promethazine 6.25 mg/5 mL. How many teaspoons will you administer?

Solutions

The health professional is frequently called upon to interpret the concentration of a solution or, on occasion, to prepare a solution. Both of these topics are discussed in this chapter.

Topics include:

Interpretation of solution concentrations expressed as:

   a percentage, W/V, V/V

   an unlabeled ratio, W/V, V/V

   a labeled ratio

Equivalent solution strength expressions

Medication problems involving interpretation of solution concentration

(Optional) Preparation of W/V and V/V solutions

(Optional) Preparation of weak solutions by dilution of strong solutions

Section 8.1   Interpreting Solution Labels

It is necessary to understand solution labeling and solution preparation because many drugs are administered in solution form. Solutions may be a mixture of liquids, or they may be a solid or solids dissolved in a liquid called a *diluent*. Solutions that are mixtures of liquids are so indicated on the label by *V/V*, which is read "volume-volume." Solutions that are a solid or solids dissolved in a diluent are so indicated on the label by *W/V*, which is read "weight-volume" or "mass-volume."

The strength of a solution can be expressed in many ways. Three commonly used methods of expressing solution strength are (1) as a percentage, (2) as an unlabeled ratio, and (3) as a labeled ratio. Each of these represents a ratio called the solution strength ratio. As noted in Section 1.1, in a ratio the numerator is the part and the denominator is the whole. In a solution strength ratio, the numerator is that part of the solution that is the drug; the denominator is the whole solution.

> When solution strength is expressed as a percentage or an unlabeled ratio, the units for measurement used in the numerator and denominator of the solution strength ratio are understood to be interpreted as follows:
>
> 1. The solution strength ratio (percentage and unlabeled ratio) of V/V solutions may be interpreted in any units for measuring volume provided the numerator and denominator are the same volume unit (Examples 1 and 4).
> 2. The solution strength ratio (percentage and unlabeled ratio) of W/V solutions is interpreted as either grams per millilitre or grains per minim (Examples 2, 3, 5, and 6). Once the label has been interpreted, equivalent expressions may be derived (Examples 3 and 5).

When solution strength is expressed as a labeled ratio, the units of measurement used in the numerator and denominator are stated on the label (Example 7).

Since a percentage is interpreted as a ratio, the solution strength percentage is really the solution strength ratio. Remember, in the solution strength ratio, the numerator is the part that is the drug and the denominator is the whole solution.

Label on container: Ethyl Alcohol 95% V/V.

EXAMPLE 1

The Ethyl Alcohol is a solution prepared by mixing liquids (V/V), and whose strength is expressed as a percentage (95%). This solution strength percentage (95%) becomes the solution strength ratio:

$$\frac{95 \text{ parts Ethyl Alcohol}}{100 \text{ parts whole solution}}$$

This is a V/V solution; therefore, the numerator and denominator may be interpreted in any units for measuring volume provided both are the same unit. In our ratio then we could be discussing litres, gallons, or millilitres—the ratio remains the same:

$$\frac{95 \text{ L Ethyl Alcohol}}{100 \text{ L solution}} \qquad \frac{95 \text{ gal Ethyl Alcohol}}{100 \text{ gal solution}} \qquad \frac{95 \text{ mL Ethyl Alcohol}}{100 \text{ mL solution}}$$

Thus 100 mL 95% (V/V) Ethyl Alcohol solution contains 95 mL Ethyl Alcohol and 5 mL distilled water to make 100 mL solution.

Label on container: sodium chloride 5%: W/V.

EXAMPLE 2

The sodium chloride solution is prepared by mixing a solid (sodium chloride) in liquid (W/V) and has a solution strength expressed as a percentage (5%). This solution strength percentage (5%) becomes the solution strength ratio $\frac{5}{100}$, which can be interpreted as grams per millilitre or grains per minim:

$$\frac{5 \text{ g sodium chloride}}{100 \text{ mL solution}} \quad \text{or} \quad \frac{\text{gr } 5 \text{ sodium chloride}}{\text{m } 100 \text{ solution}}$$

The sodium chloride 5% (W/V) solution is 5 g sodium chloride dissolved in distilled water sufficient to make 100 mL solution.

Once you have interpreted the label, you can determine equivalent expressions to the solution strength ratio.

EXAMPLE 3 | Continuing with Example 2, sodium chloride 5% W/V solution means

$$5\% \text{ W/V} = \frac{5 \text{ g sodium chloride}}{100 \text{ mL solution}}$$

Dividing by 100,

$$\frac{5 \text{ g}}{100 \text{ mL}} = \frac{0.05 \text{ g}}{1 \text{ mL}}$$

and, from Section 6.1,

$$\frac{0.05 \text{ g}}{1 \text{ mL}} = \frac{50 \text{ mg}}{1 \text{ mL}}$$

Therefore

$$\text{sodium chloride } 5\% \text{ W/V} = \frac{5 \text{ g}}{100 \text{ mL}} = \frac{50 \text{ mg}}{1 \text{ mL}}$$

or

$$\frac{50 \text{ mg sodium chloride}}{1 \text{ cm}^3 \text{ solution}}$$

Solution Strength Expressed as an Unlabeled Ratio

A second common method used to express solution strength is as an unlabeled ratio. The unlabeled ratio becomes the solution strength ratio, whose numerator is the part that is the drug and whose denominator is the whole solution.

EXAMPLE 4 | Label on container: sodium peroxide 1 : 20 V/V.

The sodium peroxide is a solution prepared by mixing liquids (V/V), and whose strength is expressed as an unlabeled ratio (1 : 20). The unlabeled ratio becomes the solution strength ratio $\frac{1}{20}$, whose numerator and denominator may be any units for measuring volume provided both are the same unit.

$$\frac{1 \text{ mL sodium peroxide}}{20 \text{ mL solution}} \quad \text{or} \quad \frac{\text{℥ 1 sodium peroxide}}{\text{℥ 20 solution}}$$

So, a sodium peroxide 1 : 20 (V/V) solution can be made by mixing 1 ounce sodium peroxide with 19 ounces distilled water to make 20 ounces solution.

Label on container: Zephiran Chloride 1 : 5000 W/V.

EXAMPLE 5

The Zephiran Chloride solution is a solution prepared by mixing a solid (Zephiran Chloride) in a liquid (W/V), and whose solution strength is expressed as an unlabeled ratio (1 : 5000). The unlabeled ratio is the solution strength ratio $\frac{1}{5000}$, which may be interpreted as grams per millilitre or grains per minim:

$$\frac{1 \text{ g Zephiran Chloride}}{5000 \text{ mL solution}}$$

or

$$\frac{\text{gr 1 Zephiran Chloride}}{\text{m } 5000 \text{ solution}}$$

The Zephiran Chloride 1 : 5000 W/V solution is 1 gram Zephiran Chloride dissolved in distilled water sufficient to make 5000 milli-litres solution.

Once the label has been interpreted, equivalent expressions may be derived.

Continuing with Example 5, Zephiran Chloride 1 : 5000 W/V solu-tion means

EXAMPLE 6

$$1 : 5000 \text{ W/V} = \frac{1 \text{ g}}{5000 \text{ mL}}$$

and, from Section 6.1,

$$\frac{1 \text{ g}}{5000 \text{ mL}} = \frac{1000 \text{ mg}}{5000 \text{ mL}}$$

and, dividing,

$$\frac{1000 \text{ mg}}{5000 \text{ mL}} = \frac{1 \text{ mg}}{5 \text{ mL}}$$

and, since

$$\frac{1 \text{ mg}}{5 \text{ mL}} = \frac{0.2 \text{ mg}}{1 \text{ mL}}$$

and

$$\frac{0.2 \text{ mg}}{1 \text{ mL}} = \frac{200 \text{ mcg}}{1 \text{ mL}}$$

EXAMPLE 6
continued

therefore,

$$\text{Zephiran Chloride } 1:5000 \text{ W/V} = \frac{1 \text{ g}}{5000 \text{ mL}}$$

$$= \frac{1 \text{ mg}}{5 \text{ mL}}$$

$$= \frac{0.2 \text{ mg}}{1 \text{ mL}}$$

$$= \frac{200 \text{ mcg Zephiran Chloride}}{1 \text{ mL solution}}$$

Solution Strength Expressed as a Labeled Ratio

A labeled-ratio solution strength is the easiest to use. Simply copy the strength printed on the container as your ratio, or derive equivalent ratios if you like.

EXAMPLE 7

Label on container: epinephrine 100 mcg/mL.

Your solution strength ratio is

$$\frac{100 \text{ mcg epinephrine}}{1 \text{ mL solution}}$$

which can be converted to

$$\frac{0.1 \text{ mg epinephrine}}{1 \text{ mL solution}}$$

Exercises 8.1

Interpret each of the following as gr/ $\mathfrak{m}$, g/100 mL, g/mL, mg/mL, and mcg/mL.

1. Novocain 1.5% W/V
2. bupivacaine 0.25% W/V
3. mannitol 20% W/V
4. Neo-Synephrine 1% W/V
5. solution 1 : 4000 W/V
6. epinephrine 1 : 10 000 W/V

Interpret each of the following in both metric and apothecaries' units.

7. solution 0.125% V/V
8. glycerin 50% V/V

9. glycerin $1:20$ V/V   10. solution $1:10$ V/V

11. solution $1:800$ V/V   12. ethyl alcohol 60% V/V

Drug Administration
Medication Problems

Section 8.2

> To solve a medication problem when the medication is in solution form, you should follow a three-step process:
>
> 1. Interpret the solution label to obtain a labeled solution strength ratio.
> 2. Convert the desired medication unit to the solution unit if necessary.
> 3. Solve the medication problem.

Remember that the solution strength on the label may be expressed as a percentage, an unlabeled ratio, or a labeled ratio.

200 mL sodium chloride 5% (W/V) has been administered over a 4-hour period. How much sodium chloride has the patient received?

EXAMPLE 1

Interpret the solution label to obtain the solution strength ratio:

$$\text{Sodium chloride 5\% W/V} = \frac{5 \text{ g sodium chloride}}{100 \text{ mL solution}} \quad (1)$$

Solve the medication problem:

$$\frac{\overset{\text{Medication on Hand}}{5 \text{ g sodium chloride}}}{100 \text{ mL solution}} = \frac{\overset{\text{Administered Medication}}{x \text{ g sodium chloride}}}{200 \text{ mL solution}} \quad (3)$$

$$100x = 1000$$
$$x = 10 \text{ g sodium chloride}$$
$$\text{in 200 mL solution}$$

The patient has received 10 g sodium chloride over the 4-hour period.

EXAMPLE 2

Desired medication: Aramine Bitartrate 6 mg, IM; medication on hand: ~~Aramine Bitartrate 1% W/V.~~

Interpret the solution label to obtain the solution strength ratio:

$$1\% \text{ W/V} = \frac{1 \text{ g}}{100 \text{ mL}} = \frac{1000 \text{ mg}}{100 \text{ mL}} = \frac{10 \text{ mg Aramine}}{1 \text{ mL solution}}$$

Solve the medication problem:

Medication on Hand		Desired Medication
$\dfrac{\text{Aramine 10 mg}}{\text{solution 1 mL}}$	$=$	$\dfrac{\text{6 mg Aramine}}{x \text{ mL solution}}$

$$10x = 6$$
$$x = 0.60 \text{ mL}$$

Administer 0.60 mL Aramine Bitartrate 1% (W/V) solution.

EXAMPLE 3

Order: epinephrine 500 mcg; medication on hand: epinephrine 1:1000 W/V.

Interpret the solution label to obtain the solution strength ratio:

$$1:1000 \text{ W/V} = \frac{1 \text{ g}}{1000 \text{ mL}} = \frac{1 \text{ mg epinephrine}}{1 \text{ mL solution}} \qquad (1)$$

Change the desired medication unit to solution unit (500 mcg = _____ mg):

$$\text{epinephrine 500 mcg} = \text{epinephrine 0.5 mg} \qquad (2)$$

Solve the medication problem:

Medication on Hand		Desired Medication
$\dfrac{\text{epinephrine 1 mg}}{\text{solution 1 mL}}$	$=$	$\dfrac{\text{0.5 mg epinephrine}}{x \text{ mL solution}}$

$$x = 0.50 \text{ mL}$$

Administer 0.50 mL epinephrine 1:1000 W/V solution.

Exercises 8.2   Solve, being sure to label and check your answers.

1. Methocarbamol 1:10 (W/V) solution is available for IM injection. The doctor asks you to administer 400 mg. How much will you administer?

2. A 1:2000 (W/V) solution is available for SC injection. To relieve migraine pain, the doctor orders 250 mcg. How much will you give?

3. Epinephrine 1:10 000 (W/V) solution is available. For asthma, the doctor prescribes 50 mcg. How much will you administer?

4. An elixir 875 mg/5 mL is available. The desired medication is 500 mg. How much will you give?

5. Acetaminophen elixir 120 mg/5 mL is available. For the alleviation of pain, administer 300 mg. How much will you give?

6. Procaine hydrochloride (Novocain) is a well-known local anesthetic. As a peripheral nerve block, administer 500 mg from a 2% (W/V) solution. How much will you give?

7. For an epidural block Novocain 1.5% (W/V) solution is used. The order is 375 mg. How much will you administer?

8. An oral suspension 0.2% (W/V) is available. Administer 10 mg. How much will you give?

9. Glycerin solution (oral) 50% (V/V) is available. The orders ask for glycerin 1.2 mL per kg body mass. How much will you administer to an 84-kg patient?

10. Mannitol solution 20% W/V is available. The orders ask for mannitol 2 g/kg body mass. How much should a 60-kg patient receive?

11. A solution 1:4000 (W/V) is used as a mouthwash. How much of the drug is contained in 100 mL?

12. A boric acid solution that contains boric acid 4% (W/V) in alcohol 70% (V/V) is available. How much boric acid and alcohol are contained in 10 mL?

13. The doctor orders epinephrine 10 mcg/kg body mass; epinephrine 1:1000 W/V is available. How much will you administer to a patient with mass of 77 lbs?

14. The doctor orders magnesium sulfate 100 mg/kg body mass; magnesium sulfate 50% W/V is available. The patient's mass is 128 lbs. How much will you administer?

15. Potassium iodide 100% W/V solution (100 g:100 mL) is available in a 30-mL dropper bottle that delivers a minim-sized drop. Administer 400 mg. How many drops will this be? (Note that the unpleasant taste can be masked by administering the drug in milk.)

16. Administer gr $\frac{1}{30}$ of 0.1% W/V solution. How much will you give?

17. Administer gr $\frac{1}{200}$ of 0.05% W/V solution. How much will you give?

Section 8.3   Preparation of W/V and V/V Solutions

In order to prepare a V/V solution or a W/V solution, you must know

1. Solution strength desired
2. The desired amount of drug
3. The total volume of solution

Given any two of these items, the third can be found by use of this formula:

$$\text{Solution strength (ratio)} = \frac{\text{Amount of drug desired}}{\text{Total volume of solution desired}}$$

Once again, the solution strength can be expressed as a percentage, an unlabeled ratio, or a labeled ratio (Section 8.1). Here we will discuss the preparation of solutions with each of the three types of solution strength.

As a Percentage   Recall that when a percentage is written in common fraction form or ratio form its denominator is always 100:

$$5\% \text{ W/V} = \frac{5 \text{ g}}{100 \text{ mL}} \qquad 0.15\% \text{ W/V} = \frac{0.15 \text{ g}}{100 \text{ mL}}$$

With this in mind, and using the formula, proceed to solve preparation problems in which the solution strength is expressed as a percentage.

Example 1 solves for the amount of drug needed when both the desired strength of the solution and the desired amount of solution are known.

How many grams sodium chloride are needed to make 800 mL of 5% sodium chloride solution?

EXAMPLE 1

$$\text{Solution strength (ratio)} = \frac{\text{Amount of drug desired}}{\text{Volume of solution}}$$

$$\frac{\text{sodium chloride 5 g}}{\text{solution 100 mL}} = \frac{x \text{ g sodium chloride}}{800 \text{ mL solution}}$$

$$100x = 4000$$

$$x = 40 \text{ g sodium chloride}$$

Sodium chloride 5% solution is prepared by dissolving 40 g sodium chloride in sufficient sterile water to make a total of 800 mL solution.

Example 2 solves for the amount solution to be prepared when given both the amount of the drug and the desired solution strength.

How much 25% medicinal zinc peroxide can be made with 60 g zinc peroxide?

EXAMPLE 2

$$\text{Solution strength (ratio)} = \frac{\text{Amount of drug desired}}{\text{Volume of solution}}$$

$$\frac{\text{zinc peroxide 25 g}}{\text{solution 100 mL}} = \frac{60 \text{ g zinc peroxide}}{x \text{ mL solution}}$$

$$25x = 6000$$

$$x = 240 \text{ mL solution}$$

Zinc peroxide 25% solution is prepared by dissolving 60 g zinc peroxide in sufficient sterile water to make 240 mL solution.

Example 3 finds the strength of solution when both the total volume of solution and the amount of drug used are known.

What percent solution is spirit of camphor made by mixing 0.3 g camphor with alcohol to make 30 mL solution?

EXAMPLE 3

$$\text{Solution strength (ratio)} = \frac{\text{Amount of drug desired}}{\text{Volume of solution}}$$

EXAMPLE 3
continued

$$\frac{\text{camphor } x \text{ g}}{\text{solution 100 mL}} = \frac{0.3 \text{ g camphor}}{30 \text{ mL solution}}$$

$$30x = 30$$

$$x = \ 1 \text{ g}$$

Spirit of camphor made by mixing 0.3 g camphor in sufficient alcohol to make 30 mL solution is a 1% spirit of camphor solution.

As an Unlabeled Ratio or as a Labeled Ratio

You should review Section 8.1 before you begin here. You will find that the method used to solve solution preparation problems here is much like that used earlier.

Example 4 solves for the amount of drug needed when both the desired strength of the solution and the desired amount of solution are known.

EXAMPLE 4

How much concentrated solution is needed to make a 1 : 10 (V/V) dilution that results in 2.5 L dilute solution?

$$\text{Solution strength (ratio)} = \frac{\text{Amount of drug desired}}{\text{Volume of solution}}$$

$$\frac{\text{conc solution 1 L}}{\text{total solution 10 L}} = \frac{x \text{ L conc solution}}{2.5 \text{ L total solution}}$$

$$10x = 2.5$$

$$x = 0.25 \text{ L concentrated solution}$$

Solution 1 : 10 (V/V) is prepared by mixing 0.25 L concentrated solution in sufficient sterile water to make 2.5 L dilute solution.

Example 5 solves for the amount of solution to be prepared when given both the amount of the drug and the desired solution strength.

EXAMPLE 5

Silver nitrate 1 g/20 mL solution is desired. How much can be made with 0.3 g silver nitrate?

$$\text{Solution strength (ratio)} = \frac{\text{Amount of drug desired}}{\text{Volume of solution}}$$

$$\frac{\text{silver nitrate 1 g}}{\text{solution 20 mL}} = \frac{0.3 \text{ g silver nitrate}}{x \text{ mL solution}}$$

$$x = 6 \text{ mL solution}$$

Silver nitrate 1 g/20 mL solution is prepared by dissolving 0.3 g silver nitrate in sufficient diluent to make 6 mL solution.

EXAMPLE 5
continued

Example 6 finds the strength of the solution when both the total volume of solution and the amount of drug used are known.

EXAMPLE 6

A solution is prepared by mixing 5 g drug in diluent sufficient to make 25 mL solution. With what solution strength will you label the container?

$$\frac{5 \text{ g}}{25 \text{ mL}} = \frac{1 \text{ g}}{5 \text{ mL}} = \frac{0.2 \text{ g}}{1 \text{ mL}} = \frac{20 \text{ g}}{100 \text{ mL}} = \frac{1000 \text{ mg}}{5 \text{ mL}} = \frac{200 \text{ mg}}{1 \text{ mL}}$$

The container may be correctly labeled in many different ways. As a percentage (see Section 8.1):

20% W/V (20 g/100 mL)

As an unlabeled ratio (see Section 8.1):

1:5 W/V (1 g/5 mL)

As a labeled ratio:

1 g/5 mL (if PO by spoon)

200 mg/mL (if by injection)

Solve, being sure to label your answers. State how to prepare the solution.

Exercises 8.3

1. How many grams of drug are needed to make 500 mL 4% (W/V) solution?

2. How many millilitres of drug are needed to make 1000 mL 10% (V/V) solution?

3. How many grams of drug are needed to make 250 mL 1:100 (W/V) solution?

4. How many millilitres of drug are needed to make 15 mL 1:200 (V/V) solution?

5. How many millilitres 5% (W/V) solution can be made with 0.2 g drug?

6. How many fluid ounces of 1% (V/V) solution can be made with f℥ $3\frac{1}{2}$ drug?

7.   How many minims of 1:20 (W/V) solution can be made from gr 12 drug?

8.   How many litres of 1:4 (V/V) solution can be made from 0.5 L drug?

9.   What percentage is a solution that has 0.2 g dissolved in water sufficient to make 10 mL solution?

10.   What percentage is a solution that has $gr \frac{1}{4}$ dissolved in diluent sufficient to make $\mathfrak{m}$ 15 solution?

11.   A solution has been prepared by dissolving 2 g drug in diluent sufficient to make 5 mL solution. Express the solution strength as

   a.   a percentage.

   b.   an unlabeled ratio.

   c.   a labeled ratio.

12.   A solution has been prepared by dissolving 0.15 g drug in diluent sufficient to make 10 mL solution. Express the solution strength as

   a.   a percentage.

   b.   an unlabeled ratio.

   c.   a labeled ratio.

13.   How is a 0.1% solution prepared using 0.5 g drug?

14.   What percent boric acid solution is made if 2 g boric acid is mixed with distilled water sufficient to make 50 mL solution?

15.   You are preparing sodium perborate 1:50 (W/V) solution for a mouthwash. How much sodium perborate is needed to prepare 200 mL solution?

16.   Sodium bicarbonate 1:20 (W/V) solution has a soothing effect on the skin. How is this prepared with 10 g sodium bicarbonate?

17.   Magnesium sulfate (Epsom Salt) 40% (W/V) solution is used as a compress or soak for the relief of inflammatory conditions. How is this prepared with 50 g Epsom Salt?

18.   For bladder irrigations a solution of 1:20 000 (W/V) silver nitrate may be used. How is this prepared with 0.5 g silver nitrate?

Preparation of Weak Solutions by Dilution of Strong Solutions

On occasion a solution is prepared by adding diluent to a strong solution to make a dilute solution. Note that *no* additional drug is added; only a diluent, such as water, is added. Instructions may read "Dilute 10 mL of 40% solution to make a 5% solution." When instructions have been executed, the amount of drug in 10 mL of 40% solution will be the *same* as that in a larger quantity of 5% solution.

> To dilute a concentrated solution, use the following relationship or formula:*
>
> $$\text{Solution strength} \times \text{volume} = \text{Solution strength} \times \text{volume}$$
> $$\text{(strong solution)} \qquad\qquad \text{(weak solution)}$$
>
> $$\text{Amount of drug in} = \text{Amount of drug in}$$
> $$\text{strong solution} \quad\ \ \text{weak solution}$$

How much 5% solution can be made from 10 mL of 40% solution?   **EXAMPLE 1**

$$
\begin{array}{cc}
\textbf{Strong Solution} & \textbf{Weak Solution} \\
(\text{strength}) \cdot (\text{volume}) & (\text{strength}) \cdot (\text{volume}) \\
(40\%) \cdot (10 \text{ mL}) = & (5\%) \cdot (x \text{ mL}) \\
(0.4) \cdot (10) = & (0.05) \cdot (x) \\
4 = & 0.05x \\
80 = & x
\end{array}
$$

---

*This is not a proportion.

EXAMPLE 1
continued

or

$$x = 80 \text{ mL } 5\% \text{ solution}$$

The 5% solution is prepared by measuring 10 mL 40% solution and adding to it diluent sufficient to make 80 mL solution.

EXAMPLE 2

How much 10% solution is needed to make 200 mL of 1.5% solution?

Concentrated Solution	Desired Solution
(strength) · (volume)	(strength) · (volume)

$$(10\%) \cdot (x \text{ mL}) = (1.5\%) \cdot (200 \text{ mL})$$
$$0.1x = (0.015) \cdot (200)$$
$$0.1x = 3$$
$$x = 30 \text{ mL of } 10\% \text{ solution}$$

The solution is prepared by taking 30 mL 10% solution and adding to it sufficient diluent to make 200 mL. You will have added 170 mL diluent (probably sterile water).

A word of caution about working with acid solutions: *Always* add the acid solution to the diluent; *never* add the diluent to the acid solution. If Example 2 discussed an acid solution, 30 mL 10% acid solution would be poured slowly into 170 mL water.

**Two Important
Observations**

1. The solution strength ratio for both the strong solution and the dilute solution must be expressed in the same units. We commonly use grams per millilitre because the percentage ratio and the unlabeled ratio are in those units.

2. The same volume unit must be used for both the strong solution and the dilute solution. That is, both must be millilitres, or both must be litres, or so on.

Exercises 8.4

Solve, being sure to label your answers. State how to prepare the solutions.

1. Prepare 150 mL 3% acetic acid solution. Acetic acid 6% solution is available.

2. Prepare 250 mL 1% Mercurochrome solution. Mercurochrome 2% solution is available.

3. Prepare 150 mL of 0.02% solution for application to wounds. A 0.1% solution is available.

4. Prepare 400 mL of 2% solution. A 5% solution is available.

5. How much 0.0025% solution can be made from 20 mL of 1% solution? The 0.0025% solution is used for wet dressing.

6. How much 0.0033% solution can be made from 10 mL of 1% solution? The 0.0033% solution is needed for a urethral irrigation.

7. How much 3% solution can be prepared from 200 mL of 5% solution?

8. How much nitrofurazone 0.1% solution can be prepared from 450 mL nitrofurazone 0.2% solution?

9. Prepare 200 mL cupric sulfate 1:4000 for use as a fungicidal agent. Cupric sulfate 5% is available.

10. Prepare 1000 mL Zephiran Chloride 1:20 000 solution. Zephiran Chloride 1:1000 is available.

Review Exercises

[8.1]

Express as gr/m, g/100 mL, g/mL, mg/mL, mcg/mL:

1. 1.5% W/V   2. 1:5000 W/V

Interpret in both metric units and in apothecaries' units:

3. 60% V/V   4. 1:5 V/V

[8.2]

Calculate dosages:

5. Give 1 g from 0.5% W/V solution.

6. Give 30 mg from 0.75% W/V.

7. Order: lidocaine 2.3 mg/kg; patient weight is 120 lbs. Available: lidocaine 4% W/V.

8. Order: epinephrine 0.3 mg/m^2, S.C.; patient body surface area is 0.85 m^2. Available: epinephrine 1:1000 W/V.

[8.3]

9. How many grams of drug are needed to make 500 mL of 5% W/V solution?

10. How many millilitres of drug are needed to prepare 10 mL of 1 : 100 V/V solution?

[8.4]

11. Tell how to prepare 10 mL of 0.5% W/V solution by diluting a 1% solution.

12. How much 1 : 4000 W/V solution can be made from 3 mL of 4% solution?

Drugs Measured in Units

The "unit" as a measurement of strength of medication is discussed in this chapter. "Drugs measured in units" is to be distinguished from "unit dosage." Drugs measured in units is a measurement system in which the strength of a drug or the amount of a drug is defined in terms of "units." Unit dosage is a drug distribution system whereby an individual, usually the pharmacist, prepares the prescribed medication for each individual patient as a separate package or "unit dosage."

Topics include:

Definition of a "unit"

Medication problems involving drugs that are measured in units and that are available as:

a solution (insulin)

a powder that must be reconstituted (penicillin)

a tablet or capsule (vitamins)

Drug Units

Many of the more recently developed drugs are measured in units. These include insulin, the penicillins, polymyxin, and some of the vitamins. A *unit* of a drug is defined in terms of the effect that specific drug has on the human body. A unit of one drug has no mass or volume relationship to a unit of a different drug.

For example, 1 unit of insulin is standardized according to its ability to lower blood glucose levels; 1 unit of penicillin G is standardized according to its antibacterial activity. There is no relationship between the unit of insulin and the unit of penicillin G. Drugs measured in units are available in many forms, five of which are:

1. solution (insulin)
2. powder that must be reconstituted (penicillin)
3. tablet or capsule form (Vitamin A)
4. ointment (polymyxin B ointment)
5. powder that is used as a powder (thrombin)

Dosage problems involving the first three types will be discussed in this chapter.

Section 9.1   ## Solution Form—Insulin

Insulin is a drug measured in units that is available in solution form. Insulin, which commonly is used in the treatment of diabetes mellitus, is a highly sensitive drug, and the highly sensitive nature of the drug mandates exact dosages. Because it is destroyed in the gastrointestinal tract when taken orally, it must be administered *parenterally,* that is, other than by way of the intestines.

As a result of the recommendation of the American Diabetes Association's Committee on the Use of Therapeutic Agents, in the United States U-80 insulin has been phased out and U-40 insulin is being phased out in favor of a single U-100 concentration for all types of insulin. In Europe U-40 insulin is the favored concentration for general use. Concentrated (U-500) insulin will continue to

be available for use in diabetes patients with daily insulin requirements greater than 200 units.

Design engineers have developed syringes marked in units to
match the solution strength of the insulin; these are called *insulin
syringes.* A 100-unit insulin syringe should be used with U-100
insulin; a 40-unit insulin syringe should be used with U-40 insulin
(see Figure 9.1).

Figure 9.1

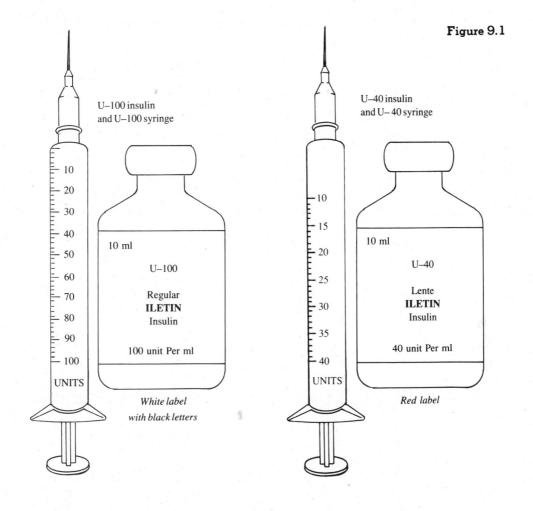

Insulin may be administered in insulin syringes or in standard
syringes. Insulin syringes, when available, always should be used,
but both methods are discussed here.

EXAMPLE 1

Order: 45 units U-100 Regular Iletin. Administer in an insulin syringe.

1. Select a bottle of insulin whose label reads *U-100 Regular Iletin*. U-100 is the strength; Regular Iletin is the type. The bottle's label is white with black lettering.

2. Select a 100-unit insulin syringe. The syringe's lettering is black.

3. Draw the U-100 Regular Iletin into the 100-unit insulin syringe up to the 45-unit level. See Figure 9.2.

Figure 9.2

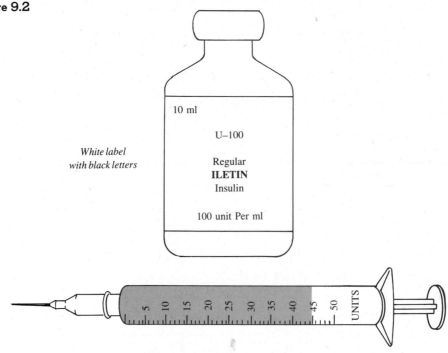

A special U–100 insulin syringe has been developed for patients who use 50 units or less of U–100 insulin.

EXAMPLE 2

Order: 30 units U-40 NPH Iletin. Administer in an insulin syringe.

1. Does the label on the container match the doctor's orders for type and strength? (Is it U-40 NPH Iletin?)

> If you administer insulin in an insulin syringe, use a simple three-step process, double-checking each step before you administer the drug.
>
> 1. Select the insulin whose label matches the doctor's order for *type* and *strength* of insulin.
>
> 2. Select an insulin syringe that matches the *strength* of the solution and the *color* of the label.
>
> 3. Draw the selected insulin into the selected syringe to the level of units ordered by the doctor.

EXAMPLE 2
continued

2. Does the insulin syringe match the strength and the label color of the ordered insulin? (Is it a 40-unit syringe? Is it red?)

3. Does the level of solution in the syringe match the number of units ordered by the doctor? (Is it at 30 units?) Figure 9.3.

Figure 9.3

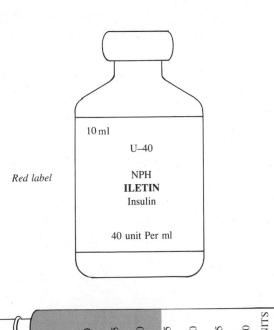

Red label

10 ml

U–40

NPH
ILETIN
Insulin

40 unit Per ml

EXAMPLE 3	Order: 65 units U-100 Regular Iletin. Administer in a 1-mL (cc) syringe.

    1.   Select a container of U-100 Regular Iletin.

    2.   U-100 insulin is defined on the label as 100 units per mL, which is the solution strength ratio. Now set up the proportion and solve for dosage.

$$\underset{\substack{\text{solution} \\ \text{strength} \\ \text{ratio}}}{\frac{100 \text{ units}}{1 \text{ mL}}} = \underset{\substack{\text{desired} \\ \text{medication}}}{\frac{65 \text{ units}}{x \text{ mL}}}$$

$$x = 0.65 \text{ mL}$$

To give 65 units U-100 Regular Iletin, draw 0.65 ml U-100 Regular Iletin into a 1-mL (cc) syringe and administer the solution (Figure 9.4).

Figure 9.4

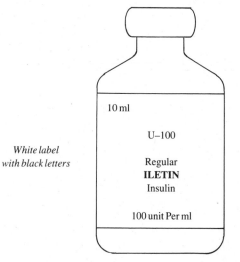

White label
with black letters

10 ml

U–100

Regular
ILETIN
Insulin

100 unit Per ml

EXAMPLE 4	Order: 310 units U-500 Regular Iletin. Administer in a 1-mL (cc) syringe.

    1.   Select a container of U-500 Regular Iletin.

You can administer insulin in a standard 1-mL (cc) tuberculin syringe when a matching insulin syringe is not available. Again, double-check each step.

1. Select the insulin preparation whose *type* and *strength* matches that ordered by the doctor.

2. Using the solution strength ratio printed on the label of the selected insulin, set up a proportion to solve for the volume of solution to be administered in the 1-mL (cc) syringe.

2. U-500 insulin is defined on the label as 500 units per mL, which is its solution strength ratio.

EXAMPLE 4
continued

$$\frac{\overset{\text{solution strength ratio}}{500 \text{ units}}}{1 \text{ mL}} = \frac{\overset{\text{desired medication}}{310 \text{ units}}}{x \text{ mL}}$$

$$x = 0.62 \text{ mL}$$

To give 310 units U-500 Regular Iletin, draw 0.62 mL U-500 Regular Iletin into a 1-mL (cc) syringe and administer the solution (see Figure 9.5).

Figure 9.5

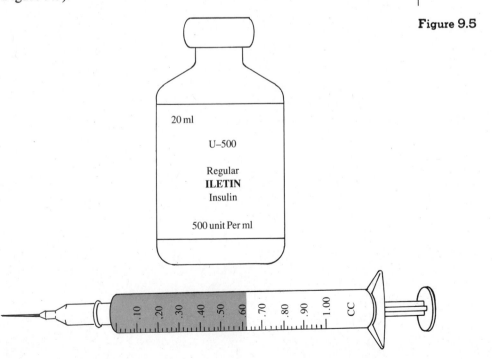

Exercises 9.1   Discuss and give the correct syringe settings.

1. Order: 60 units U-100 Regular Iletin in a 100-unit insulin syringe.
2. Order: 40 units U-100 Regular Iletin in a 100-unit insulin syringe.
3. Order: 60 units U-100 Regular Iletin in a tuberculin (1 mL) syringe.
4. Order: 45 units U-100 Regular Iletin in a tuberculin (1 mL) syringe.
5. Order: 25 units U-40 isophane insulin Suspension in a 40-unit insulin syringe.
6. Order: 20 units U-40 NPH Iletin in a 40-unit insulin syringe.
7. Order: 250 units U-500 Regular Iletin in tuberculin syringe.
8. Order: 200 units U-500 Regular Iletin in tuberculin syringe.

Section 9.2   # Powder Form That Must Be Reconstituted

The penicillins in powder form are examples of those drugs that need to be reconstituted to be administered intramuscularly (IM) or intravenously (IV). To reconstitute penicillin, dissolve the dry powder in a diluent (liquid) to form a solution. In powder form, the penicillins generally are stable for several years at room temperature; in reconstituted form, the penicillins deteriorate rapidly. Recommended maximum storage time can be found on the vial label. Directions for reconstituting the penicillins also are found on the vial labels, and these allow some choice of concentration.

To mix and use a drug that needs to be reconstituted, follow this procedure:

1. Select a solution concentration from those choices given on the label, keeping two things in mind:
   a. The total volume for intramuscular injection should not exceed 4 mL. For adults, volumes up to 2 mL are best; volumes larger than 2 mL should be divided and administered in two separate syringes in two separate locations.
   b. The number of units of the drug the doctor wants administered.

2. Prepare the solution for the selected concentration according to the label's directions.

3. Label the bottle of prepared solution, indicating its concentration and the date of preparation. (Adhesive tape makes a good label.)

4. Use the selected solution-concentration ratio as the medication-on-hand ratio to solve medication problems.

Use the information on this label to work Examples 1 and 2.

Buffered Potassium Penicillin G for Injection
5 000 000 units
Preparation of Solution

Add diluent	Concentration
18.2 mL	250 000 units/mL
8.2 mL	500 000 units/mL
3.2 mL	1 000 000 units/mL

Orders: patient A—400 000 units penicillin G potassium, IM.

EXAMPLE 1

1. Select the 18.2 mL diluent solution concentration diluent to give a concentration of 250 000 units/mL. (This enables you to administer 400 000 units in the desirable volume range.)

2. Prepare the solution. Add 18.2 mL diluent to the powder in the vial. Dissolve completely.

3. Label the vial:

   250 000 units/mL and the date

4. Solve the medication problem:

$$\frac{\text{medication on hand}}{\text{250 000 units}} = \frac{\text{desired medication}}{\text{400 000 units}}$$

$$\frac{250\ 000\ \text{units}}{1\ \text{mL}} = \frac{400\ 000\ \text{units}}{x\ \text{mL}}$$

$$x = 1.6\ \text{mL}$$

Administer to patient A 1.6 mL penicillin G potassium (250 000 units/mL), IM.

EXAMPLE 2 | Orders: patient B—initial dose, 1 500 000 units penicillin G potassium, IM.

This is the same type of penicillin ordered for patient A in Example 1. The problem then is: 1.6 mL (400 000 units) was used for patient A from a vial that originally contained 18.2 mL (5 000 000 units). Although there is enough penicillin in the vial, should it be used? To decide, calculate the dosage.

$$\underset{\substack{\text{medication} \\ \text{on hand}}}{\frac{250\ 000 \text{ units}}{1 \text{ mL}}} = \underset{\substack{\text{desired} \\ \text{medication}}}{\frac{1\ 500\ 000 \text{ units}}{x \text{ mL}}}$$

$$x = 6 \text{ ml penicillin G potassium}$$
$$(250\ 000 \text{ units/mL})$$

6 mL exceeds the allowable volume; therefore, you should prepare a new solution, using a new vial and following the procedure.

1. Select the 3.2 ml diluent solution concentration to give a concentration of 1 000 000 units/mL.

2. Prepare the solution. Add 3.2 ml diluent to the powder in the vial. Dissolve completely.

3. Label the vial:

      1 000 000 units/mL and the date

4. Solve the medication problem.

$$\underset{\substack{\text{medication} \\ \text{on hand}}}{\frac{1\ 000\ 000 \text{ units}}{1 \text{ mL}}} = \underset{\substack{\text{desired} \\ \text{medication}}}{\frac{1\ 500\ 000 \text{ units}}{x \text{ mL}}}$$

$$x = 1.5 \text{ mL}$$

Administer to patient B 1.5 mL penicillin G potassium (1 000 000 units/mL), IM.

Exercises 9.2 | 1. Give 50 000 units reconstituted sodium penicillin G from a multiple-dose vial, in which 10 mL contain 1 000 000 units.

2. From a vial labeled *heparin sodium, 20 000 units/mL* give 8000 units heparin sodium.

3. From a vial labeled *heparin sodium, 15 000 units/mL* give 20 000 units heparin sodium.

4. From a vial labeled *ACTH 20 units/mL* give ACTH 12.5 units, QID.

5. From a vial of reconstituted bacitracin labeled *5000 units bacitracin/mL* administer 1000 units/kg body mass in two divided doses. The patient's mass is 15 lbs. Show all your work.

6. Order: bacitracin 10 000 units every 6 hours; available: reconstituted bacitracin 5000 units/mL and reconstituted bacitracin 1000 units/ml. Which would you use? Show your work.

7. Order: penicillin G potassium 50 000 units/kg body mass, as 4 divided doses. Available: reconstituted penicillin G 250 000 units/mL. Patient's mass is 50 lbs.

8. Order: penicillin G procaine 1.2 million units; available: penicillin G procaine 600 000 units/mL.

9. Order: bleomycin 10 units/m^2 body surface; available: reconstituted bleomycin 10 units/mL. Patient has 1.69 m^2 body surface.

Tablets and Capsule Forms

Section 9.3

Bicillin tablets, penicillin G potassium tablets, and vitamin E capsules are examples of drugs measured in units that are available in tablet or capsule form. You will find the calculations simple.

Order: Bicillin 500 000 units QID; medication on hand: Bicillin tablets 200 000 units scored.

EXAMPLE 1

$$\frac{\overset{\text{medication}}{\underset{\text{on hand}}{}}}{} \qquad \overset{\text{desired}}{\underset{\text{medication}}{}}$$

$$\frac{200\ 000\ \text{units}}{1\ \text{tablet}} = \frac{500\ 000\ \text{units}}{x\ \text{tablets}}$$

$$x = 2\frac{1}{2}\ \text{tabs}$$

Administer $2\frac{1}{2}$ tablets Bicillin (200 000 units), QID.

Exercises 9.3   1.   Order: vitamin E 100 units; available: vitamin E 50 units/capsule and vitamin E 100 units/capsule.

2.   Order: oleovitamin A 9000 units/day; available: oleovitamin A 10 000 units/capsule.

3.   Order: 50 000 units penicillin G potassium/kg body mass total daily, in four equally divided doses; available: penicillin G potassium 100 000 units/tablet, 200 000 units/tablet, 250 000 units/tablet, and 400 000 units/tablet. Patient's mass is 88 lbs.

4.   Order: 55 000 units penicillin G potassium/kg body mass/day, administered PO in four equally divided doses; available: penicillin G potassium as in Exercise 3. Patient's mass is $50\frac{1}{2}$ lbs.

Review Exercises   [9.1]

1.   Administer 70 units of Regular Iletin from U-100 Regular Iletin:

   a.   using a 100-unit insulin syringe

   b.   using a 1-mL tuberculin syringe

2.   Administer 25 units of U-40 insulin using a 40-unit insulin syringe.

3.   Administer 270 units of U-500 insulin using a tuberculin (1 mL) syringe.

[9.2]

4.   Order: bacitracin 900 units/kg, IM, daily as 2 equally divided doses; patient weight 2.5 kg. Available: reconstituted bacitracin 5000 units /mL.

[9.3]

5.   Order: penicillin G benzathine 60 000 units/kg, PO, daily as 4 equally divided doses; patient weight 44 lbs. Available: penicillin G benzathine 200 000 units/tablet (scored).

Intravenous Drug Administration

This chapter contains detailed discussion of the mathematics involved in the administration of drugs by Intravenous (IV) method. When medication is delivered by Intravenous method, a large volume of solution, which contains the drug(s), is administered continuously over an extended period of time. Because of the extended period of time, the system must be checked many times and, if necessary, adjustments made. Calculations require understanding of rate problems to make the original setting of the system and to make the adjustments over time.

Topics include:

Different types of Intravenous (IV) delivery equipment

Medication problems involving one or more of the following:

calculation of amount of drug needed by the individual patient

interpretation of solution labels

calculation of amount of solution to be delivered

calculation of "rate of flow" attained by use of the various IV delivery systems

adjustment of "rate of flow" to attain timely delivery of medication

all other complicating factors found in other medication problems in this text.

Intravenous Delivery System

Intravenous administration (IV) delivers drugs over an extended period of time by the drop method. That is, the drugs enter the bloodstream a drop at a time. The advantages of intravenous delivery are twofold.

1. The drug is delivered directly to the bloodstream, quickly reaching that part of the body requiring medication.

2. The drug is delivered at a steady or constant rate over an extended period of time, which makes it easier for the body to assimilate and which is a more effective use of a drug. In contrast, drugs administered orally (PO), such as tablets and syrups, can deliver an initial excess of a drug and then a lack of it.

The type of IV delivery system determines the size of the drop. For example, four different IV delivery systems are constructed to deliver 10 drops/mL, 15 drops/mL, 20 drops/mL, and 60 drops/mL. The 10 drops/mL, 15 drops/mL, and 20 drops/mL systems are regulated by counting drops per minute and adjusting the flow valve to the desired rate. Therefore, the rate-of-flow calculations must be made per minute. The 60 drops/mL is called the *microdropper delivery system,* and it delivers such a small drop that actual counting is almost impossible. In fact, a counting mechanism has been developed so that a dial may be set on the desired rate of flow.

To calculate the delivery rate of flow, you must know

1. the total amount of drug to be delivered
2. the total time of delivery
3. the type of IV delivery system being used

There are two methods of calculation. The first uses two separate proportion problems.

1. Determine the millilitres per minute.
2. Apply the dosage per minute to the available IV system.

EXAMPLE 1

Order: 150 mL dextrose 5% to be delivered in 1 hour via 10 drops/mL delivery system. The dextrose 5% is the liquid in the delivery bottle.

EXAMPLE 1
continued

The important numbers are 150 mL and the delivery time of 1 hour, which is converted to minutes (60 minutes). Now calculate the millilitres per minute:

$$\frac{150 \text{ mL}}{60 \text{ min}} = \frac{x \text{ mL}}{1 \text{ min}}$$

$$x = 2.5 \text{ mL/min}$$

Deliver 2.5 mL/min to deliver 150 mL/h dextrose 5%.

Via the 10 drops/mL delivery system:

$$\frac{\text{delivery system ratio}}{} \quad \frac{\text{desired medication per minute}}{}$$

$$\frac{10 \text{ drops}}{1 \text{ mL}} = \frac{x \text{ drops}}{2.5 \text{ mL dextrose } 5\%}$$

$$x = 25 \text{ drops}$$

25 drops per minute via the 10 drops/mL delivery system will deliver 2.5 mL/min or 150 mL/h dextrose 5%.

EXAMPLE 2

Order: 150 mL dextrose 5% to be delivered in 1 hour via the 15 drops/ml delivery system. From Example 1, the millilitres per minute is 2.5.

$$\frac{15 \text{ drops}}{1 \text{ mL}} = \frac{x \text{ drops}}{2.5 \text{ mL dextrose } 5\%}$$

$$x = 37.5 \text{ drops}$$

38 drops per minute via the 15 drops/mL delivery system will deliver 2.5 mL/min or 150 mL/h dextrose 5%.

EXAMPLE 3

Order: 150 mL dextrose 5% to be delivered in 1 hour via the 20 drops/mL delivery system. From Example 1, the millilitres per minute is 2.5.

$$\frac{20 \text{ drops}}{1 \text{ mL}} = \frac{x \text{ drops}}{2.5 \text{ mL dextrose } 5\%}$$

EXAMPLE 3
continued

$$x = 50 \text{ drops}$$

50 drops per minute via the 20 drops/mL delivery system will deliver 2.5 mL/min or 150 mL/h dextrose 5%.

EXAMPLE 4

Order: 150 mL dextrose 5% to be delivered in 1 hour via the microdropper (60 drops/mL) delivery system.

<table>
<tr><td>delivery
system
ratio</td><td></td><td>desired
medication
per minute</td></tr>
</table>

$$\frac{60 \text{ drops}}{1 \text{ mL}} = \frac{x \text{ drops}}{2.5 \text{ mL dextrose } 5\%}$$

$$x = 150 \text{ drops}$$

150 drops per minute via the microdropper delivery system will deliver 2.5 mL/min or 150 mL/h dextrose 5%.

It is important to note that when you use the microdropper system, the desired millilitres per hour (150) is the same number as the drops per minute (150).

Solving Examples 1 through 4 involved two steps: first, finding the rate in millilitres per minute and, second, finding the rate in drops per minute based on the specific delivery system. A careful study of the mathematics involved shows both may be accomplished with one formula if the following is known:

1. The total millilitres of drug to be delivered
2. The total number of minutes for delivery
3. The type of delivery system

With this information, apply the formula:

$$\frac{\left(\begin{array}{l}\text{Drops per millilitre of the}\\\text{delivery system}\end{array}\right) \cdot \left(\begin{array}{l}\text{Total millilitres drug}\\\text{to be delivered}\end{array}\right)}{\text{Total minutes for delivery}}$$

$$= \text{Drops per minute}$$

Examples 5 through 8 use the same order used in Examples 1 through 4: 150 mL dextrose 5% delivered in 1 hour.

Via the 10 drops/mL delivery system.

EXAMPLE 5

$$\frac{(10 \text{ drops/mL}) \cdot (150 \text{ mL})}{60 \text{ min}} = 25 \text{ drops/min}$$

25 drops per minute delivers 150 mL/h via the 10 drops/mL delivery system.

Via the 15 drops/mL delivery system.

EXAMPLE 6

$$\frac{(15 \text{ drops/mL}) \cdot (150 \text{ mL})}{60 \text{ min}} = 37.5 \text{ drops/min}$$

38 drops per minute delivers 150 mL/h via the 15 drop/mL delivery system.

Via the 20 drop/mL delivery system.

EXAMPLE 7

$$\frac{(20 \text{ drops/mL}) \cdot (150 \text{ mL})}{60 \text{ min}} = 50 \text{ drops/min}$$

50 drops per minute delivers 150 mL/h via the 20 drop/mL delivery system.

Via the microdropper (60 drops/mL) delivery system.

EXAMPLE 8

$$\frac{(60 \text{ drops/mL}) \cdot (150 \text{ mL})}{60 \text{ min}} = 150 \text{ drops/min}$$

150 drops per minute delivers 150 mL/h via the microdropper delivery system.

Again, note that when you use the microdropper system, the desired millilitres per hour (150) is the same number as the drops per minute (150). Calculations can be made for times greater than 1 hour.

EXAMPLE 9   Order: 1 L sodium lactate 5% in 6 hours via microdropper.

$$1 \text{ L} = 1000 \text{ mL}$$
$$6 \text{ h} = 360 \text{ min}$$
$$\text{microdropper} = 60 \text{ drops/mL}$$

Solve using the formula:

$$\frac{(60 \text{ drops/mL}) \cdot (1000 \text{ mL})}{360 \text{ min}} = 166.\overline{6} \text{ drops/min}$$

167 drops per minute delivers 1 L sodium lactate 5% in 6 h via the microdropper delivery system.

EXAMPLE 10   In Example 9 (1 L sodium lactate 5% in 6 h via microdropper), assume you have come on duty 2 h after delivery has started. 280 mL sodium lactate 5% has been delivered. How will you adjust the rate?

$$1000 \text{ mL} - 280 \text{ mL} = 720 \text{ mL left to be delivered}$$
$$6 \text{ h} - 2 \text{ h} = 240 \text{ min (4 h)}$$
$$\text{microdropper} = 60 \text{ drops/mL}$$

Solve using the formula:

$$\frac{(60 \text{ drops/mL}) \cdot (720 \text{ mL})}{240 \text{ min}} = 180 \text{ drops/min}$$

180 drops per minute via the microdropper will deliver 720 mL sodium lactate 5% in 4 h.

IV calculations can be a part of a larger problem, as in Example 11.

EXAMPLE 11   A 130-lb patient is to receive IV, over a period of 24 h, dextran 40 in 10% solution equal to 2 grams dextran 40 per kilogram body mass via the microdropper delivery system.

1. Change 130 lbs to kg (Section 7.2):

$$\frac{130 \text{ lbs}}{x \text{ kg}} = \frac{2.2 \text{ lbs}}{1 \text{ kg}}$$
$$x = 59 \text{ kg}$$
$$130 \text{ lbs} = 59 \text{ kg}$$

EXAMPLE 11
continued

2. Find the total grams of dextran 40 needed at 2 g/kg:

$$\frac{2\text{ g}}{1\text{ kg}} = \frac{x\text{ g}}{59\text{ kg}}$$

$$x = 118\text{ g dextran 40}$$

3. Calculate the amount of 10% dextran 40 solution that contains 118 g (Section 8.2, Example 2). Remember that 10% equals $\frac{10}{100}$.

$$\frac{10\text{ g dextran 40}}{100\text{ mL soln}} = \frac{118\text{ g dextran 40}}{x\text{ mL soln}}$$

$$x = 1180\text{ mL 10\% dextran 40 solution}$$

4. Solve the IV problem using the formula.

$$\text{total mL required} = 1180\text{ mL}$$
$$\text{total time in minutes} = 24 \times 60 = 1440\text{ min}$$
$$\text{delivery system} = 60\text{ drops/mL}$$
$$\frac{(60\text{ drops/mL}) \cdot (1180\text{ mL})}{1440\text{ min}} = 49\text{ drops/min}$$

49 drops per minute via the microdropper will deliver 1180 mL 10% dextran 40 solution in 24 h; 1180 mL 10% dextran 40 contains 118 g dextran 40; 118 g dextran 40 is 2 g/kg body mass.

Exercises

1. Order: $\frac{1}{6}$ molar sodium lactate 275 mL/h, IV.
    a. Via 10 drops/mL delivery system
    b. Via 15 drops/mL delivery system
    c. Via 20 drops/mL delivery system
    d. Via microdropper delivery system

2. Order: 5% dextrose 175 mL/h, IV.
    a. Via 10 drops/mL delivery system
    b. Via 15 drops/mL delivery system
    c. Via 20 drops/mL delivery system
    d. Via microdropper delivery system

3. Order: 5% Saline Soln 300 mL over 6 h, IV.
    a. Via 10 drops/mL IV system
    b. Via 15 drops/mL IV system
    c. Via 20 drops/mL IV system
    d. Via microdropper IV system

4. Order: Glucose 5% soln 600 mL over 8 h, IV.

    a. Via 10 drops/mL IV system
    b. Via 15 drops/mL IV system

    c. Via 20 drops/mL IV system
    d. Via microdropper IV system

5. Order: Glucose 5% 1000 mL/8 h, IV via 15 drops/mL delivery system. After 3h, 450 mL has been delivered; calculate a new rate to finish on time.

6. Order: dextrose 5% 1 L/6 h, IV via microdropper. After 2h, 200 mL has been delivered; calculate a new rate to finish on time.

7. A 150-lb patient with severe acidosis is to be given isotonic sodium lactate 40 mL/kg body mass over 14h. Calculate for

    a. total volume to be administered.

    b. rate using microdropper delivery system.

8. A 100-lb patient is to receive IV, over a period of 24 h, dextran 70 in 6% soln equal to 1.2 g dextran 70/kg body mass. Calculate for

    a. total volume to be administered.

    b. rate using microdropper delivery system.

9. 2 L dextrose 10% soln is to be administered IV at the rate of 500 mg dextrose/kg body mass per hour. The patient has a mass of 130 lbs. Calculate for

    a. total time to administer.

    b. rate using microdropper delivery system.

10. A 150-lb patient is to receive IV, over a period of 24 h, dextran 75 in 6% soln equal to 600 mg dextran 75/kg body mass. Calculate for

    a. total volume to be administered.

    b. rate using microdropper delivery system.

Pediatric Dosages

Determination of correct pediatric dosage is not easy; optimally, dosage is based on clinical observation of the sick child. The search for the "best rule" that reflects research findings has continued for decades. The five "rules" included in this chapter represent the various stages of this search up to, and including, the most recent and most accurate rule.

Topics include:

Rules for determining pediatric dosages: Rule for dose per square metre of body surface

    Clark's body surface rule

    Fried's rule (infants)

    Young's rule

    Clark's weight

Use of the West Nomogram

Dosage Determination

Infants and children do not receive the same amount of medication as do adults. Although a doctor prescribes the medication, a nurse has the moral and legal responsibility to know what constitutes proper pediatric dosages of common drugs. Determination of "proper child dosage" is not always easy. It would be best if each child dosage were based on clinical observation, but insufficient data make this impossible. Therefore, various formulas have been devised to assist in making dosage determinations. Of the many available, here are five that might be considered general guides.

Based on Body Surface Area Expressed in Square Metres

1. A formula that emancipates the child dose from the adult dose

$$\text{Body surface area (in m}^2) \times \text{Dose/m}^2$$
$$= \text{Approximate child dose}$$

2. Clark's body surface rule

$$\frac{\text{Body surface area of child (in m}^2)}{1.73} \times \text{Adult dose}$$
$$= \text{Approximate child dose}$$

Based on Age

3. Fried's rule (infants younger than one year)

$$\frac{\text{Age in months}}{150} \times \text{Adult dose}$$
$$= \text{Approximate infant dose}$$

4. Young's rule

$$\frac{\text{Age in years}}{\text{Age in years} + 12} \times \text{Adult dose}$$
$$= \text{Approximate child dose}$$

Based on Weight

5. Clark's weight rule

$$\frac{\text{Weight in pounds}}{150} \times \text{Adult dose}$$
$$= \text{Approximate child dose}$$

Each of the five methods has its limitations, some more serious than others. At best, each gives only an approximate dosage. Formula 1, based on body surface area, has been found the most satisfactory in clinical observations of children; this formula has the added advantage of being appropriate through almost the entire life span of an individual. The other formulas are presented for historical perspective and understanding. Formulas 3, 4, and 5 are used for "over the counter" drugs since patient weight or age are obtainable by the lay public. Formulas based on age are least satisfactory. None of the formulas is satisfactory for the early neonatal period of premature and term infants.

Take a few minutes to study the West Nomogram in the Appendix. You can see that body surface area is based on both the height and the weight of the individual.

Order: epinephrine (1 : 1000), 300 mcg/m^2 body surface; patient: a child of height 35 inches and weight 28 lbs. Using Formula 1, calculate the proper dose.

EXAMPLE 1

1. Using the West Nomogram, determine the body surface.

    35 inches and 28 lbs = 0.56 m^2 body surface

2. Using Formula 1, determine the dosage.

    $$\frac{\text{Body surface}}{\text{area (m}^2)} \times \text{Dose/m}^2 = \frac{\text{Approximate}}{\text{child dose}}$$

    $(0.56 \text{ m}^2) \times (300 \text{ mcg/m}^2) = 168 \text{ mcg epinephrine}$

3. Administer 168 mg epinephrine from epinephrine (1 : 1000).

    1 : 1000 = 1 g/1000 mL = 1 mg/1 mL

    168 mcg = 0.168 mg

    $$\frac{\overset{\text{medication}}{\underset{\text{on hand}}{}} }{}$$

    $$\frac{\text{1 mg epinephrine}}{\text{1 mL soln}} = \frac{\text{0.168 mg epinephrine}}{x \text{ mL soln}}$$

    $$x = 0.17 \text{ mL}$$

Administer 0.17 mL epinephrine (1 : 1000) to give 300 mcg/m^2 body surface to a child who is 35 inches tall and who weighs 28 lbs.

EXAMPLE 2

Using Clark's body surface rule, calculate the proper dose of Gantrisin for a child with 0.63 m² body surface. The average adult Gantrisin dose is 4 g.

$$\frac{\text{Body surface area of child (m}^2)}{1.73} \times \text{Adult dose} = \begin{array}{c}\text{Approximate}\\\text{child dose}\end{array}$$

$$\frac{0.63 \text{ m}^2}{1.73} \quad \times \quad 4 \text{ g} \quad = \quad 1.46 \text{ g}$$

According to Clark's body surface rule, administering 1.46 g Gantrisin to a child with 0.63 m² body surface is approximately equivalent to administering 4 g to an adult.

EXAMPLE 3

Using Fried's rule, calculate the proper dose of Gantrisin for a 10-month-old infant. The average adult dose is 4 g.

$$\frac{\text{Age in months}}{150} \times \text{Adult dose} = \begin{array}{c}\text{Approximate}\\\text{infant dose}\end{array}$$

$$\frac{10}{150} \quad \times \quad 4 \text{ g} \quad = \quad 0.27 \text{ g}$$

According to Fried's rule, administering 0.27 g Gantrisin to a 10-month-old infant is approximately equivalent to administering 4 g Gantrisin to an adult.

EXAMPLE 4

Using Young's rule, calculate the proper dose of aspirin for a 6-year-old child. The average adult aspirin dose is 600 mg.

$$\frac{\text{Age in years}}{\text{Age in years} + 12} \times \text{Adult dose} = \begin{array}{c}\text{Approximate}\\\text{child dose}\end{array}$$

$$\frac{6}{6 + 12} \quad \times \quad 600 \text{ mg} \quad =$$

$$\frac{6}{18} \quad \times \quad 600 \text{ mg} \quad = \quad 200 \text{ mg}$$

According to Young's rule, administering 200 mg aspirin to a 6-year-old child is approximately equivalent to administering 600 mg aspirin to an adult.

Using Clark's weight rule, calculate the correct dose of atropine sulfate for a 40-lb child if the average adult dose is 0.4 mg.

EXAMPLE 5

$$\frac{\text{Weight in pounds}}{150} \times \text{Adult dose} = \frac{\text{Approximate}}{\text{child dose}}$$

$$\frac{40}{150} \qquad \times \quad 0.4 \text{ mg} \quad = 0.11 \text{ mg}$$

According to Clark's weight rule, administering 0.11 mg atropine sulfate to a 40-lb child is approximately equivalent to administering 0.4 mg to an adult.

Calculate the approximate child dosage. State the formula used.   **Exercises**

1. Calculate the dosage of Nembutal for a 5-month-old infant if the adult Nembutal dosage is 90 mg.

2. Calculate the dosage of penicillin G for a 7-month-old infant if the adult dosage is 500 000 units.

3. Calculate the dosage of tetracycline for an 8-year-old child if the adult dosage is 250 mg.

4. Calculate the dosage of Demerol for a $3\frac{1}{2}$-year-old child if the adult dosage is 120 mg.

5. Calculate the dosage of gentamicin for a 28-lb child if the adult dosage is 80 mg.

6. Calculate the dosage of Cleocin for a 52-lb child if the adult dosage is 200 mg.

7. Calculate the penicillin G dosage for a child with 0.69 m^2 body surface. The average adult penicillin G dose is 300 000 units.

8. Calculate the dosage of chloral hydrate for a child who is 44 inches tall and who weighs 45 lbs. The usual adult dose is 250 mg.

9. Calculate the dosage of sulfisoxazole for a child who is 56 inches tall and who weighs 73 lbs. The order is for sulfisoxazole 2 g/m^2 body surface.

10. Calculate the dosage of hydrocodone bitartrate for a child who is 62 inches tall and who weighs 84 lbs. The order is for hydrocodone bitartrate 20 mg/m^2 body surface administered PO in 4 equally divided doses. Hydrocodone bitartrate is available as an oral liquid 5 mg/5 mL.

Burn Patients: Estimating Percentage of Burn

This chapter presents two methods for determining percentage of body surface burned.

Topics include:

Rule of Nine

A second method that adjusts for patient age and that has accurate percentage measures for the various body surface areas. This second method is considered to be the more accurate one.

Estimating Percentage of Burn

Burn patient care and survival prediction depend on many variables, but three primary variables are percent of body burned, degree of the burn, and patient vital signs. Decisions concerning vital signs and degree of a burn are based on knowledge you will obtain in your more advanced courses. Percent of body burned is a mathematical decision and is considered in this chapter.

One method that has long been used for estimating percent of body burned is the Rule of Nine (study Figure 12.1). By the Rule of Nine, if the total skin surface on the right arm is burned, we say 9% of the body surface is burned; if one-half of the total skin surface on the left arm is burned, $\frac{1}{2}$ of 9%, or $4\frac{1}{2}\%$, of the body surface is burned; if one-fourth of the skin surface on the right leg is burned, $\frac{1}{4}$ of 18%, or $4\frac{1}{2}\%$, of the body surface is burned. If a patient has these burn areas, the total burn area is estimated to be $9\% + 4\frac{1}{2}\% + 4\frac{1}{2}\%$, or 18%, of the total body surface. Estimates of other burned areas work in the same manner. Thus, the Rule of Nine provides a rough estimate of percentage of body burned but makes no provision for difference in body size due to age.

Rule of Nine

Figure 12.1

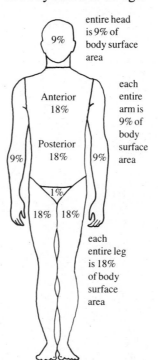

entire head is 9% of body surface area

9%

Anterior 18%

Posterior 18%

each entire arm is 9% of body surface area

9%

9%

1%

18%   18%

each entire leg is 18% of body surface area

Rule of Nine

Source: J. C. Scherer, 1982.

Age-based Method for Estimating Percentage of Burn

A *more recently developed method* for estimating the percent of body surface burned (Figure 12.2) makes two major improvements: *More accurate percentages are assigned to the various surface areas and adjustments are made for difference in body size.* You should study the figures and charts to understand both methods, but pay particular attention to the second method, which is considered to be the more accurate one. In the chart titled "Percent of Areas Affected by Growth," note that during growth years the percentage of body surface representing the head decreases; by contrast, during growth years the percentage of body surface representing the thigh and leg increases.

Figure 12.2

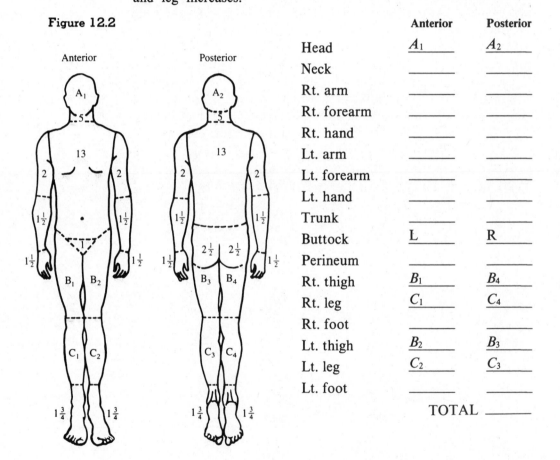

	Anterior	Posterior
Head	A_1	A_2
Neck		
Rt. arm		
Rt. forearm		
Rt. hand		
Lt. arm		
Lt. forearm		
Lt. hand		
Trunk		
Buttock	L	R
Perineum		
Rt. thigh	B_1	B_4
Rt. leg	C_1	C_4
Rt. foot		
Lt. thigh	B_2	B_3
Lt. leg	C_2	C_3
Lt. foot		
	TOTAL	

Percent of Areas Affected by Growth*

	0	1	5	10	15	Adult
A = ½ Head	9½	8½	6½	5½	4½	3½
B = ½ One thigh	2¾	3¼	4	4¼	4½	4¾
C = ½ One leg	2½	2½	2¾	3	3¼	3½

For assistance with calculations review Section 3.3.   **Adult**

	Anterior	Posterior	EXAMPLE 1
Head	A_1 1.17	A_2 3.5	
Neck	2.5	3.75	
Rt. arm			
Rt. forearm			
Rt. hand			
Lt. arm	2.0	2.0	
Lt. forearm	1.12	1.12	
Lt. hand	0.75	0.75	
Trunk	4.33	6.5	
Buttock	L 1.88	R	
Perineum			
Rt. thigh	B_1	B_4	
Rt. leg	C_1	C_4	
Rt. foot			
Lt. thigh	B_2 1.19	B_3 1.19	
Lt. leg	C_2	C_3	
Lt. foot			
TOTAL	33.8%		

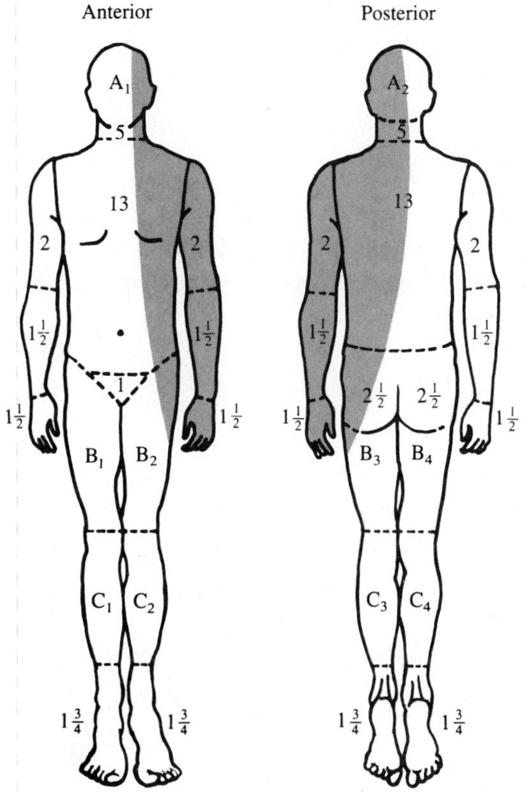

*Burn Evaluation Chart for estimation of percent body burns according to age
Source: Crozer Chester Medical Center, Upland, PA 19013.

EXAMPLE 1
continued

Percent of Areas Affected by Growth

	0	1	5	10	15	Adult
$A = \frac{1}{2}$ Head	9½	8½	6½	5½	4½	3½
$B = \frac{1}{2}$ One thigh	2¾	3¼	4	4¼	4½	4¾
$C = \frac{1}{2}$ One leg	2½	2½	2¾	3	3¼	3½

Anterior

Head A_1: $\frac{1}{3}$ of $3\frac{1}{2}\%$

$\quad \frac{1}{3}$ of 3.5% = 1.17%

Neck: $\frac{1}{2}$ of 5% = 2.5%

Left arm: 2%

Left forearm: $\frac{3}{4}$ of $1\frac{1}{2}\%$

$\quad$ 0.75 of 1.5% = 1.12%

Left hand: $\frac{1}{2}$ of $1\frac{1}{2}\%$

$\quad$ 0.5 of 1.5% = 0.75%

Trunk: $\frac{1}{3}$ of 13% = 4.33%

Left thigh B_2: $\frac{1}{4}$ of $4\frac{3}{4}\%$

$\quad$ 0.25 of 4.75% = 1.19%

Posterior

Head A_2: $3\frac{1}{2}\%$ = 3.5%

Neck: $\frac{3}{4}$ of 5%

$\quad$ 0.75 of 5% = 3.75%

Left arm: 2%

Left forearm: $\frac{3}{4}$ of $1\frac{1}{2}\%$ = 1.12%

Left Hand: $\frac{1}{2}$ of $1\frac{1}{2}\%$ = 0.75 %

Trunk: $\frac{1}{2}$ of 13% = 6.5%

Buttock L (left): $\frac{3}{4}$ of $2\frac{1}{2}\%$

$\quad$ 0.75 of 2.5% = 1.88%

Right thigh B_3: $\frac{1}{4}$ of $4\frac{3}{4}\%$

$\quad \frac{1}{4}$ of 4.75% = 1.19%

A clear plastic metric ruler will help estimate burn area.

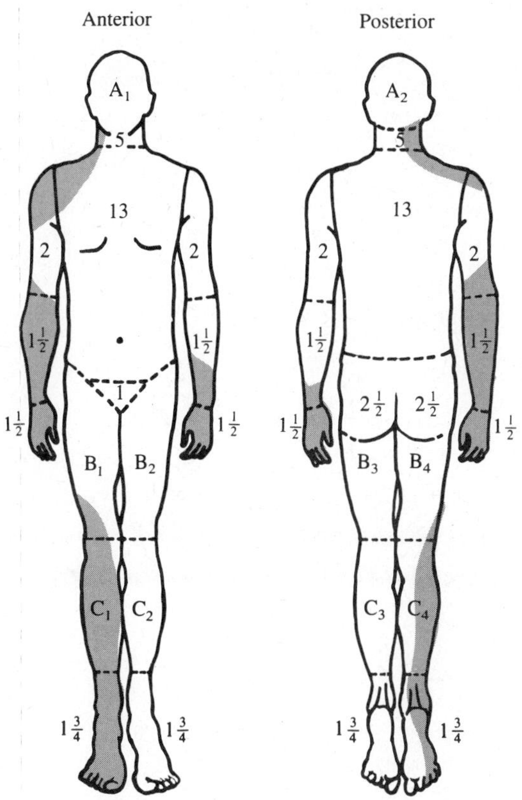

Anterior   Posterior

	Anterior	Posterior	EXAMPLE 2
Head	A_1	A_2	
Neck	1.25	1.25	
Rt. arm	1.00	0.67	
Rt. forearm	1.50	1.5	
Rt. hand	1.50	1.5	
Lt. arm			
Lt. forearm	0.75	0.5	
Lt. hand	1.50	1.5	
Trunk	1.30	1.1	
Buttock	L	R	
Perineum			
Rt. thigh	B_1 1.00	B_4 0.8	
Rt. leg	C_1 2.06	C_4 1.38	
Rt. foot	1.75	0.88	
Lt. thigh	B_2	B_3	
Lt. leg	C_2	C_3	
Lt. foot			
	TOTAL	24.7%	

Percent of Areas Affected by Growth						
	0	1	5	10	15	Adult
A = ½ Head	9½	8½	6½	5½	4½	3½
B = ½ One thigh	2¾	3¼	4	4¼	4½	4¾
C = ½ One leg	2½	2½	2¾	3	3¼	3½

EXAMPLE 2 continued	Anterior	Posterior

Anterior

Neck: $\frac{1}{4}$ of 5% = 1.25%

Right arm: $\frac{1}{2}$ of 2% = 1%

Right forearm: $1\frac{1}{2}$% = 1.5%

Right hand: $1\frac{1}{2}$% = 1.5%

Left forearm: $\frac{1}{2}$ of $1\frac{1}{2}$% = 0.75%

Left hand: $1\frac{1}{2}$% = 1.5%

Trunk: $\frac{1}{10}$ of 13% = 1.3%

Right thigh B_1: $\frac{1}{4}$ of 4% = 1%

Right leg C_1: $\frac{3}{4}$ of $2\frac{3}{4}$% = 2.06%

Right Foot: $1\frac{3}{4}$% = 1.75%

Posterior

Neck: $\frac{1}{4}$ of 5% = 1.25%

Right arm: $\frac{1}{3}$ of 2% = 0.67%

Right forearm: $1\frac{1}{2}$% = 1.5%

Right hand: $1\frac{1}{2}$% = 1.5%

Left forearm: $\frac{1}{3}$ of $1\frac{1}{2}$% = 0.5%

Left hand: $1\frac{1}{2}$% = 1.5%

Trunk: $\frac{1}{12}$ of 13% = 1.1%

Right thigh B_4: $\frac{1}{5}$ of 4% = 0.8%

Right leg C_4: $\frac{1}{2}$ of $2\frac{3}{4}$% = 1.38%

Right foot: $\frac{1}{2}$ of $1\frac{3}{4}$% = 0.88%

Exercises   1. One-year-old child; estimate percent of burn.

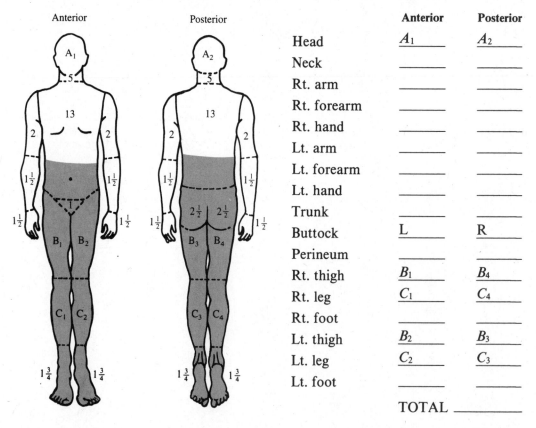

	Anterior	Posterior
Head	A_1	A_2
Neck		
Rt. arm		
Rt. forearm		
Rt. hand		
Lt. arm		
Lt. forearm		
Lt. hand		
Trunk		
Buttock	L	R
Perineum		
Rt. thigh	B_1	B_4
Rt. leg	C_1	C_4
Rt. foot		
Lt. thigh	B_2	B_3
Lt. leg	C_2	C_3
Lt. foot		
TOTAL		

2. Ten-year-old child; estimate percent of burn.

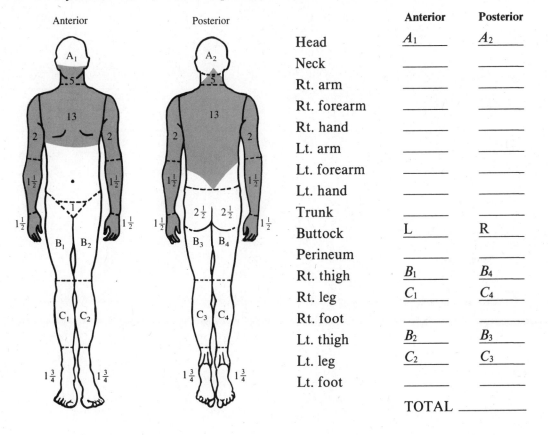

	Anterior	Posterior
Head	A_1	A_2
Neck		
Rt. arm		
Rt. forearm		
Rt. hand		
Lt. arm		
Lt. forearm		
Lt. hand		
Trunk		
Buttock	L	R
Perineum		
Rt. thigh	B_1	B_4
Rt. leg	C_1	C_4
Rt. foot		
Lt. thigh	B_2	B_3
Lt. leg	C_2	C_3
Lt. foot		
TOTAL		

3. Adult; estimate percent of burn.

Anterior   Posterior

	Anterior	Posterior
Head	A_1	A_2
Neck		
Rt. arm		
Rt. forearm		
Rt. hand		
Lt. arm		
Lt. forearm		
Lt. hand		
Trunk		
Buttock	L	R
Perineum		
Rt. thigh	B_1	B_4
Rt. leg	C_1	C_4
Rt. foot		
Lt. thigh	B_2	B_3
Lt. leg	C_2	C_3
Lt. foot		

TOTAL _____

Anterior figure labels: A_1, .5, 13, 2, 2, $1\frac{1}{2}$, $1\frac{1}{2}$, 1, $1\frac{1}{2}$, $1\frac{1}{2}$, B_1, B_2, C_1, C_2, $1\frac{3}{4}$, $1\frac{3}{4}$

Posterior figure labels: A_2, .5, 13, 2, 2, $1\frac{1}{2}$, $1\frac{1}{2}$, $1\frac{1}{2}$, $1\frac{1}{2}$, $2\frac{1}{2}$, $2\frac{1}{2}$, B_3, B_4, C_3, C_4, $1\frac{3}{4}$, $1\frac{3}{4}$

4. Five-year-old child; estimate percent of burn.

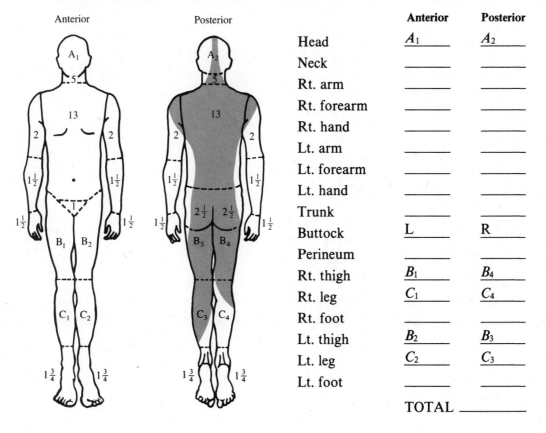

	Anterior	Posterior
Head	A_1	A_2
Neck		
Rt. arm		
Rt. forearm		
Rt. hand		
Lt. arm		
Lt. forearm		
Lt. hand		
Trunk		
Buttock	L	R
Perineum		
Rt. thigh	B_1	B_4
Rt. leg	C_1	C_4
Rt. foot		
Lt. thigh	B_2	B_3
Lt. leg	C_2	C_3
Lt. foot		
TOTAL		

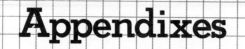

Appendixes

Whole Number Addition and Multiplication Facts

Addition

+	2	3	4	5	6	7	8	9	10	11	12
2	4										
3	5	6									
4	6	7	8								
5	7	8	9	10							
6	8	9	10	11	12						
7	9	10	11	12	13	14					
8	10	11	12	13	14	15	16				
9	11	12	13	14	15	16	17	18			
10	12	13	14	15	16	17	18	19	20		
11	13	14	15	16	17	18	19	20	21	22	
12	14	15	16	17	18	19	20	21	22	23	24

Multiplication

×	2	3	4	5	6	7	8	9	10	11	12
2	4										
3	6	9									
4	8	12	16								
5	10	15	20	25							
6	12	18	24	30	36						
7	14	21	28	35	42	49					
8	16	24	32	40	48	56	64				
9	18	27	36	45	54	63	72	81			
10	20	30	40	50	60	70	80	90	100		
11	22	33	44	55	66	77	88	99	110	121	
12	24	36	48	60	72	84	96	108	120	132	144

Figure A.1

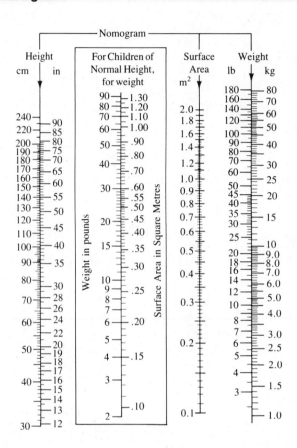

The surface area is indicated where a straight line connecting the height and weight intersects the surface area column. Or if the patient is roughly of average size, use the weight alone (enclosed area).

Modified from data of E. Boyd by C. D. West; from H. C. Shirkey, "Drug Therapy" in *Textbook of Pediatrics,* 9th ed. (W. E. Nelson and V. C. Vaughn III, eds.), W. B. Saunders Co., Philadelphia, 1964.

Temperature: Celsius, Fahrenheit, and Kelvin Scales

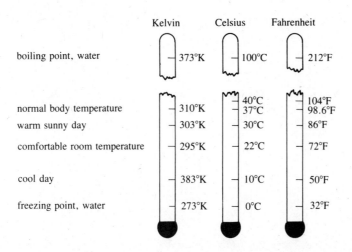

Figure A.2

The kelvin scale is used to measure temperatures in scientific work. One degree kelvin is equal to 1 degree Celsius. The Celsius and Fahrenheit scales are used to measure air temperatures, cooking temperatures, and body temperatures. The Celsius scale is the official temperature scale for ordinary use in the United States.

$$\text{Celsius} = C \qquad \text{Fahrenheit} = F \qquad \text{Kelvin} = K$$

$$^\circ C = \frac{(^\circ F - 32)}{1.8} \qquad ^\circ F = (1.8 \cdot {}^\circ C) + 32 \qquad ^\circ K = {}^\circ C + 273$$

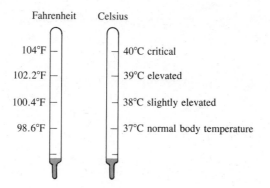

Figure A.3

Figure A.4

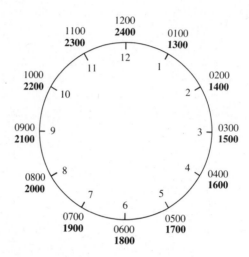

The lightface numbers represent 1:00 AM through noon. The bold-face numbers represent 1:00 PM through midnight. To use the 24-hour clock in speaking, follow the table.

12-Hour Clock	24-Hour Clock	Spoken 24-Hour Time
1 AM	0100	*zero one hundred,* or *o one hundred*
7:30 AM	0730	*zero seven thirty,* or *o seven thirty*
8:05 AM	0805	*zero eight o five,* or *o eight o five*
10:15 AM	1015	*ten fifteen*
2:45 PM	1445	*fourteen forty-five*
6:13 PM	1813	*eighteen thirteen*
midnight	2400	*twenty-four hundred*

Metric Prefixes, Symbols, and Values

Prefix	SI Symbol	Value
tera	T	$1\ 000\ 000\ 000\ 000 = 10^{12}$
giga	G	$1\ 000\ 000\ 000 = 10^{9}$
mega	M	$1\ 000\ 000 = 10^{6}$
kilo	k	$1\ 000 = 10^{3}$
hecto	h	$100 = 10^{2}$
deka	da	$10 = 10^{1}$
		reference unit = 1
deci	d	$0.1 = 10^{-1}\left(\frac{1}{10}\right)$
centi	c	$0.01 = 10^{-2}$
milli	m	$0.001 = 10^{-3}$
micro	μ	$0.000\ 001 = 10^{-6}$
nano	n	$0.000\ 000\ 001 = 10^{-9}$
pico	p	$0.000\ 000\ 000\ 001 = 10^{-12}$
femto	f	$0.000\ 000\ 000\ 000\ 001 = 10^{-15}$
atto	a	$0.000\ 000\ 000\ 000\ 000\ 001 = 10^{-18}$

Metric–Apothecary–U.S. Customary Equivalents

The metre is a generous yard. 1 metre is approximately 3.4 inches longer than 1 yard.

$$\frac{1 \text{ m}}{39.4 \text{ in}}$$

or

$$\frac{1 \text{ yd}}{0.9 \text{ m}} \qquad \text{(approximations)}$$

Litre—Quart   The litre is a generous quart.

$$1 \text{ quart } = 32 \text{ fluid ounces}$$

$$\frac{\text{f} \mathcal{Z} \, 1}{30 \text{ mL}} = \frac{\text{f} \mathcal{Z} \, 32}{x \text{ mL}} \qquad \text{(Section 6.1)}$$

$$x = 960 \text{ mL (32 fluid ounces, or 1 quart)}$$

Therefore, 1 liter (1000 millilitres) is 40 millilitres more than 1 quart. 40 millilitres equals 2 tablespoons plus 2 teaspoons.

$$\frac{1 \text{ qt}}{0.96 \text{ L}}$$

or

$$\frac{1 \text{ L}}{1.04 \text{ qt}} \qquad \text{(approximations)}$$

Conventional Cup—Metric Cup

$$1 \text{ qt} = 960 \text{ mL}$$

$$\frac{1}{4} \text{ qt} = 1 \text{ cup (conventional)}$$

$$\frac{1}{4} \cdot 960 \text{ mL} = 240 \text{ mL in 1 conventional cup}$$

$$1 \text{ L} = 1000 \text{ mL}$$

$$\frac{1}{4} \text{ L} = 1 \text{ cup (metric)}$$

$$\frac{1}{4} \cdot 1000 \text{ mL} = 250 \text{ mL in 1 metric cup}$$

Therefore, the metric cup (used in European recipes) is 10 milli-litres more than the conventional cup. 10 millilitres is 2 teaspoons (Section 7.1), which probably is not enough difference to matter in most recipes.

Kilometre—Mile

The ratio

$$\frac{1 \text{ km}}{0.6 \text{ mi}}$$

is a good one to remember. A quick way to change kilometres to miles is to mark off (drop) one position first and then multiply by 6.

> 40 km/h
> 4.0 drop one position
> ×6 multiply by 6
> 24 mi/h

Metric Ton— Conventional Ton

A metric ton, sometimes called a *long ton,* is 2240 pounds; the conventional ton is 2000 pounds.

Symbols and Abbreviations

@	at		NPO	nothing by mouth
a	before		OD	right eye
aa	of each, an equal quantity		OM	each morning; also *o.m.*
AC	before meals		ON	each night; also *o.n.*
ad lib	freely as desired		OS	left eye
AM	before noon		oz,$\mathfrak{Z}$	ounce
amt	amount		$\bar{p}$	after; also *p*
aq	water; also H_2O		$\overline{pc}$	after meals; also *pc*
ASAP	as soon as possible		PO	by mouth
BID	twice a day; also *b.i.d.*		PM	after noon
BP	blood pressure		PRN	when necessary
$\bar{c}$	with		$\bar{q}$	each; also *q*
°C	degree Celsius		QD	daily; also *q.d.*
caps	capsules		QID	four times daily; also *q.i.d.*
CBC	complete blood count			
cm	centimetre		QOD	every other day; also *q.o.d.*
cm^3	cubic centimetre; also *cc*		q h	every hour
comp	compound		q 2 h	every two hours
conc	concentrate; concentrated; concentration		q 3 h	every three hours
			q 4 h	every four hours
cr	cream		QS	as much as required
dil	dilute		qt	quart
dr,$\mathfrak{Z}$	dram		r	rectal
D5W	5% dextrose in water; also 5 DW		s	without
			SC	subcutaneous; also *s.c.*
elix	elixir			
°F	degree Fahrenheit		Sig	label
g	gram		soln	solution
gr	grain		$\overline{ss}$	one half; also *ss*
gtt	drops		stat	immediately
h	hour; also *hr*		subl	sublingual
hs	bedtime		syr	syrup

IM	intramuscular; also *i.m.*	t	topical
IV	intravenous; also *i.v.*	T	temperature
IVPB	IV piggyback	tab	tablet
kg	kilogram	tbsp	tablespoon
L	litre	TPR	temperature-pulse-respiration
lb	pound	tsp	teaspoon
lcm	least common multiple	TID	three times daily; also *t.i.d.*
m	metre	U	unit
m^2	square metre	ung	ointment
mcg,		↓	decreased
μg	microgram	↑	increased
mg	milligram	<	is less than
ℳ	minim	>	is greater than
mm	millimetre	≈	approximately equal to
μm	micrometre, micron	=	equal to
		≠	not equal to

ampere A unit used to measure the strength of electric current.

angstrom (Å) A unit used to measure the length of light waves; 1 angstrom equals 0.000 000 000 1 metre.

body mass The total mass of the human body, measured in kilograms or pounds.

body surface area The total surface area of the skin that covers the human body, measured in square metres (m^2).

candela A unit used to measure luminous intensity.

centimetre (cm) One hundredth metre. A unit of length commonly used to express the linear measure of larger parts of the human body; 2.54 centimetres equals 1 inch.

cubic centimetre (cm^3) A cube each edge of which is 1 centimetre long. The volume of 1 cubic centimetre equals 1 millilitre.

denominator In a common fraction that number below the line or fraction bar, representing the whole amount under discussion.

desired medication The type and quantity of drug prescribed by a doctor.

diluent A liquid, such as water or alcohol, that is used to dissolve a solid or solids. A diluent also may be a liquid added to a solution in order to dilute or weaken the solution.

divided dose The total dosage of a drug (usually over 24 hours), which is to be separated into a stated number of equal parts. Each part then is administered at given intervals of time, such as BID, TID, or QID.

dividend In division, the number that is to be separated into equal parts; the numerator of a common fraction.

divisor In division, the number of equal parts into which a quantity is to be separated; the number used to divide by; the denominator of a common fraction.

elixir A sweetened medicinal preparation, usually alcoholic, that is administered orally.

equivalent Has the same value as; is equal to; may be substituted for.

excrete To give off or expel.

grain (gr) The basic unit of mass in the apothecaries' system of measurement; 1 grain equals $\frac{1}{1000}$ avoirdupois pound. In ancient times a grain was considered equal to a grain of wheat taken from the center of a field.

gram (g) One thousandth kilogram. A unit of mass commonly used to express the mass of a drug.

identity property of multiplication The product of any given number and 1 is the given number; $a \cdot 1 = 1 \cdot a = a$.

intramuscular (IM) The administration of drugs via injection into a muscle or muscle tissue.

intravenous (IV) The administration of drugs via injection directly into a vein.

isotonic A salt solution that has the same osmotic pressure as the blood.

kelvin (K) A unit for measuring temperature; 1 degree kelvin equals 1 degree Celsius.

kilogram (kg) The basic unit for measuring mass (commonly a patient's mass) in the metric system; 1 kilogram equals approximately 2.2 avoirdupois pounds.

litre (L) The basic unit for measuring volume in the metric system; 1 litre equals 1.04 quarts. The litre is equal in volume to 1 cubic decimetre.

medication on hand The prescribed drug in the form available in the supply cupboard or from the pharmacist.

metre (m) The basic unit for measuring length in the metric system; 1 metre equals 39.4 inches. The metre is the length equal to 1 650 763.73 wavelengths, in vacuum, of the radiation corresponding to the transition between the levels $2p_{10}$ and $5d_5$ of the krypton—86 atom.

metric ton 1000 kilograms; also called a *long ton.*

microgram (mcg, μg) One millionth gram. A unit of mass commonly used to express the mass of drugs.

micron (μm) One millionth metre. A unit of length used to express the linear measure of those parts of the human body that are not visible to the naked eye; also called *micrometre.*

milligram (mg) One thousandth gram; one millionth kilogram. A unit of mass commonly used to express the mass of drugs.

millilitre (mL) One thousandth litre. A unit of volume commonly used to express the volume of drugs; 1 millilitre is equivalent in volume to 1 cubic centimetre; 1 teaspoon is approximately equal to 5 millilitres.

millimetre (mm) One thousandth metre. A unit of length commonly used to express the linear measure of small but visible parts of the human body; 25.4 millimetres equals 1 inch.

minim (♍) The basic unit of volume in the apothecaries' system of measurement; roughly the size of a drop of water. 16.23 minims equal 1 millilitre.

mole An amount of a substance that contains 6.02×10^{23} molecules.

neonatal Relating to an infant during its first month of life.

numerator In a common fraction, that number above the line or fraction bar, which represents the part under discussion.

oral Having to do with the mouth; by mouth.

parenteral Administering drugs by some means other than the intestines (oral)—IV, IM, or SC.

peripheral Pertaining to the outer surface.

postoperative Following surgery.

proportion Two equal ratios.

quotient In division, the answer.

radian A unit used to measure a plane angle.

ratio A comparison of two quantities, written $a : b$ or $\frac{a}{b}$.

solution (V/V) A solution prepared by mixing two or more liquids.

solution (W/V) A solution prepared by dissolving a solid, or solids, in a liquid; see *diluent*.

square metre (m^2) A square that is 1 metre on each side.

standard pressure 1 atmosphere pressure; 760 millimetres.

standard temperature Zero degrees Celsius (0° C).

steradian A unit used to measure a solid angle.

subcutaneous (SC) The administration of drugs via injection beneath the skin.

sublingual Under the tongue.

suspension A uniform dispersion of the fine particles of a solid in a liquid; many oral medications are suspensions.

topical A particular part of the body; a local application of a drug.

vial A multiple-dose container used for injectables.

References

American Society for Testing and Materials. *Metric Practice Guide.* E380-72. 1916 Race St., Philadelphia, Pa., 1972.

American Society of Hospital Pharmacists. *American Hospital Formulary Service.* 4630 Montgomery Ave., Bethesda, Md. 1984.

Frost, D. V.; Helgren, F. J.; and Sokol, L. F. *Metric Handbook for Hospitals.* U.S. Metric Association, Sugarloaf Star Route, Boulder, Colo., 1975.

Gray, Henry. *Anatomy of the Human Body.* 28th ed. Edited by Charles Mayo Goss. Philadelphia: Lea and Febiger, 1969.

Guralnik, David B., ed. *Webster's New World Dictionary of the American Language.* 2d college ed. New York: World, 1970.

Shirkey, Harry C. *Pediatric Dosage Handbook.* American Pharmaceutical Association, 2215 Constitution Ave., N.W., Washington, D.C., 1973.

U.S. Department of Commerce, National Bureau of Standards. *The International Bureau of Weights and Measures 1875-1975.* NBS Special Publication 420. Washington, D.C.: U.S. Government Printing Office, 1975.

Scherer, J. C. *Introductory Medical-Surgical Nursing.* 3rd ed. Philadelphia, J. B. Lippincott, 1982.

Answers to Reviews and Odd-numbered Exercises

1.

3.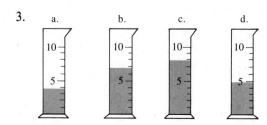

5. a. $\dfrac{5}{6}$

    b. $\dfrac{3}{4}$

    c. $\dfrac{7}{10}$

7.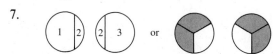

1. 2, 3	15. prime number (83 · 1)
3. 3	17. 2 · 2 · 2 · 3
5. none	19. 2 · 2 · 3 · 3 · 5 · 7 · 11
7. 3, 5	21. 2 · 31
9. 2, 3, 5, 7, 11, 13, 17, 19, 23, 29, 31, 37	23. prime number (37 · 1)
	25. 23 · 29
11. 3 · 5	27. 3 · 3
13. 2 · 3 · 3	29. 3 · 29

Exercises 1.2

1. $\dfrac{3}{6}$

3. $\dfrac{80}{20}$

5. $\dfrac{10}{5}$

7. $\dfrac{63}{336}$

9. $\dfrac{3}{48}$

11. $\dfrac{3}{7}$

Exercises 1.3

13. $\dfrac{20}{21}$

17. $\dfrac{7}{10}$

15. $\dfrac{421}{483}$
The numerator and denom-
inator have no prime
factors in common.

19. $\dfrac{3}{29}$

Exercises 1.4   1. 6

9. 90

3. 60

11. 210

5. 120

13. 60

7. 6

15. 48

Exercises 1.5

1. $\dfrac{7}{24}$

7. $1\dfrac{17}{20}$

3. $\dfrac{23}{336}$

9. $\dfrac{31}{105}$

5. $\dfrac{5}{84}$

Exercises 1.6

1. $3\dfrac{3}{4}$

15. $\dfrac{50}{3}$

3. $8\dfrac{1}{2}$

17. $\dfrac{1}{2}$

5. $2\dfrac{1}{22}$

19. $11\dfrac{29}{95}$

7. $1\dfrac{2}{3}$

21. $9\dfrac{1}{8}$

9. $\dfrac{41}{5}$

23. $4\dfrac{1}{8}$

11. $\dfrac{65}{9}$

25. $\dfrac{73}{75}$

13. $\dfrac{10}{3}$

27. $16\dfrac{19}{24}$ ounces

Exercises 1.7

1. $\dfrac{1}{8}$

7. $4\dfrac{5}{8}$ or $\dfrac{37}{8}$

3. $\dfrac{1}{143}$

9. $2\dfrac{2}{3}$ or $\dfrac{8}{3}$

5. $3\dfrac{3}{8}$ or $\dfrac{27}{8}$

11. $\dfrac{2}{225}$

13. $\dfrac{24}{35}$

15. $\dfrac{9}{1000}$

17. $\dfrac{17}{225}$

19. $3\dfrac{41}{60}$ or $\dfrac{221}{60}$

21. $3\dfrac{3}{4}$ tablets

23. $2\dfrac{1}{7}$ times

25. 4 ounces

27. 2 times

1. $\dfrac{1}{2}$

3. $\dfrac{1}{2}$

5. $2\dfrac{3}{35}$ or $\dfrac{73}{35}$

7. 8

9. $\dfrac{1}{25}$

Exercises 1.8

1. $\dfrac{1}{3} < \dfrac{1}{2}$

3. $\dfrac{3}{8} > \dfrac{3}{11}$

5. $\dfrac{1}{120} > \dfrac{1}{125}$

7. $\dfrac{5}{8} < \dfrac{7}{11}$

9. $\dfrac{4}{7} < \dfrac{2}{3}$

11. $\dfrac{1}{3} < \dfrac{7}{18} < \dfrac{5}{12}$

13. $\dfrac{6}{13} < \dfrac{5}{8} < \dfrac{7}{11}$

15. more nearly $\dfrac{33}{100}$

Exercises 1.9

1.
a. $\dfrac{1}{4}$

b. $\dfrac{2}{3}$

c. $\dfrac{1}{2}$

d. $\dfrac{3}{4}$

2. a. $\dfrac{5}{18}$

b. $\dfrac{13}{18}$

3. a. by 5
b. not by 2, 3, or 5
c. by 3 and 5
d. by 2 and 3

4. a. $3 \cdot 17 \left(3\overline{\smash)51}^{\,17} \right)$

b. $13 \cdot 13 \left(13\overline{\smash)169}^{\,13} \right)$

c. $2 \cdot 3 \cdot 19 \left(2\overline{\smash)114}^{\,3\overline{\smash)57}^{\,19}} \right)$

Review

$$\begin{array}{r} 5 \\ 5\overline{)\ 25} \\ 2\overline{)\ 50} \\ 2\overline{)\ 100} \\ 2\overline{)\ 200} \end{array}$$

d. $2 \cdot 2 \cdot 2 \cdot 5 \cdot 5$

5.   a. $\dfrac{15}{65}$

 b. $\dfrac{20}{160}$

 c. $\dfrac{30}{5}$

 d. $\dfrac{49}{35}$

6.   a. $\dfrac{3 \cdot 3 \cdot 3 \cdot \cancel{5}}{2 \cdot \cancel{5} \cdot 11 \cdot 13} = \dfrac{27}{286}$

 b. $\dfrac{\cancel{2} \cdot 19}{2 \cdot \cancel{2} \cdot 13} = \dfrac{19}{26}$

 c. $\dfrac{2 \cdot 19}{899} = \dfrac{38}{899}$

 Since 899 is not divisible by 2 or by 19, the fraction will not reduce.

 d. $\dfrac{\cancel{2} \cdot 3 \cdot \cancel{5}}{2 \cdot \cancel{2} \cdot \cancel{5} \cdot 5} = \dfrac{3}{10}$

7.   a. $3 \cdot 3 \cdot 5 \cdot 7 = 315$
 b. $2 \cdot 2 \cdot 3 \cdot 3 = 36$
 c. $2 \cdot 3 \cdot 5 \cdot 7 \cdot 11 = 2310$
 d. $2 \cdot 3 \cdot 5 \cdot 7 = 210$

8.   a. $5\dfrac{3}{4}$

 b. $1\dfrac{8}{9}$

 c. $4\dfrac{3}{4}$

 d. $3\dfrac{3}{4}$

9.   a. $\dfrac{57}{8}$

 b. $\dfrac{19}{4}$

 c. $\dfrac{85}{8}$

 d. $\dfrac{29}{6}$

10.   a. $1\dfrac{187}{315}$ or $\dfrac{502}{315}$

 b. $\dfrac{1}{600}$

 c. $3\dfrac{25}{72}$ or $\dfrac{241}{72}$

 d. $8\dfrac{3}{5}$ or $\dfrac{43}{5}$

11.   a. $\dfrac{4}{3}$ or $1\dfrac{1}{3}$

 b. $\dfrac{13}{108}$

 c. $1\dfrac{7}{12}$ or $\dfrac{19}{12}$

 d. $\dfrac{1}{60}$

12. $3\dfrac{5}{6}$ times or about 4 times

13. $\dfrac{3}{4}$

14. $7\dfrac{7}{16}$ or $\dfrac{119}{16}$

15.   a. $1\dfrac{1}{5}$ or $\dfrac{6}{5}$

 b. $2\dfrac{14}{15}$ or $\dfrac{44}{15}$

 c. $1\dfrac{1}{3}$ or $\dfrac{4}{3}$

 d. $\dfrac{1}{450}$

16.   a. $\dfrac{1}{300} < \dfrac{1}{100}$

      b. $\dfrac{7}{9} > \dfrac{10}{13}$

      c. $\dfrac{13}{17} = \dfrac{52}{68}$

      d. $\dfrac{1}{100} > \dfrac{1}{200}$

Chapter 2

Exercises 2.1

1.   10
3. $10 \cdot 10 \cdot 10 \cdot 10$
5. $\dfrac{1}{10 \cdot 10 \cdot 10}$
7.   10
9.   10 000

11. $\dfrac{1}{1000}$
13. 10^4
15. 10^1
17. $\dfrac{1}{10^3}$

Exercises 2.2

1. $4(10^2) + 8(10^1) + 2(1)$
3. $3(10^3) + 5(1)$
5. $6(1) + 7\left(\dfrac{1}{10}\right) + 4\left(\dfrac{1}{10^2}\right)$
7.   6.7
9.   0.000 067
11.   0.06
13.   0.001 65
15.   23.000 048

17.   Eight and three-millionths
19.   Six and seventy-four hundredths
21.   Ten thousand seventy-eight hundred thousandths
23.   One hundred sixty-five hundred thousandths
25.   One-hundredth
27.   One ten-thousandth

Exercises 2.3

1.   18.09
3.   8.15
5.   4.702
7.   100.053
9.   13.127
11.   40.418

13.   16.5955
15.   9.0909
17.   10.642
19.   24.438
21.   6.7 grams

Exercises 2.4

1.   0.2346
3.   0.85
5.   0.168
7.   0.0036

9.   276.624
11.   45 800
13.   0.000 135
15.   11.085

17. 7008
19. 0.006 854
21. 4.5 milligrams daily;
     31.5 milligrams weekly

23. 0.128 millilitre
25. 3.71 milligrams per dose;
     14.84 milligrams daily

Exercises 2.5

1. 0.3
3. 2.0
5. 5.6
7. 0.33
9. 7.39
11. 10.00
13. 0.333
15. 4.587
17. 13.000
19. 0.213
21. 4.688

23. 3.2
25. 0.312
27. 0.031
29. 1.756
31. 209.070
33. 0.453
35. 61.364
37. 0.001
39. 0.3 part
41. 0.6 times

Exercises 2.6

1. $0.02 > 0.0025$
3. $0.885 < 1.00$
5. $0.040 = 0.04$
7. $0.105 > 0.0105$
9. $\dfrac{1}{1000} = 0.001 < 0.100$

11. $0.003 < 0.03 < 0.047$
     $< 0.157 < 0.3 < 0.33$
13. No. $0.0035 < 0.0040$;
     you gave too much.

Exercises 2.7

1. $\dfrac{3}{5}$
3. $\dfrac{3}{5000}$
5. $\dfrac{27}{40}$
7. $6\dfrac{1}{2}$
9. $\dfrac{1}{10}$
11. 0.75

13. 0.143
15. 1.6
17. 0.168
19. $0.\overline{2}$
21. $\dfrac{1}{2} = 0.50$
23. $0.3 < \dfrac{1}{3}$
25. $\dfrac{7}{6} < 1.2$

Review

1. a. $10 \cdot 10 \cdot 10 \cdot 10 \cdot 10$
    b. $\dfrac{1}{10 \cdot 10 \cdot 10}$
2. a. 10^5
    b. $\dfrac{1}{10^4}$
3. a. 1000
    b. $\dfrac{1}{100}$

4. a. $3(10^2) + 4(10) + 8(1)$

    b. $7(1) + 4\left(\dfrac{1}{10^2}\right) + 8\left(\dfrac{1}{10^4}\right)$

5. a. 108.013

    b. 2.0060

6. a. nine-thousandths

    b. two hundred and one hundred nine ten-thousandths

7. a. 31.013

    b. 17.607

    c. 8.491

    d. 8.756

8. a. 0.153

    b. 0.001 44

    c. 60 500

    d. 8.9

9. 2.1 millilitres

10. 0.51 kilogram

11. a. 3.904

    b. 3.90

    c. 3.9

    d. 4

12. a. 0.210

    b. 4.762

    c. 0.059

    d. 0.008

13. 0.21, or about 0.2 part

14. 4.8, or about 5 times

15. a. $0.04 > 0.0045$

    b. $6.06 = 6.060$

16. a. $\dfrac{13}{125}$

    b. $70\dfrac{2}{25}$ or $\dfrac{1752}{25}$

17. a. $0.\overline{45}$

    b. 0.68

18. a. $0.03 < \dfrac{1}{30}$

    b. $\dfrac{25}{4} > 6.24$

    c. $0.125 = \dfrac{1}{8}$

    d. $0.33 < \dfrac{1}{3}$

Chapter 3

Exercises 3.1

1. $\dfrac{9}{50}$

3. $\dfrac{17}{200}$

5. $\dfrac{9}{2}$

7. $\dfrac{1}{50}$

9. $\dfrac{3}{10\ 000}$

11. 0.04

13. 0.0394

15. 0.00125

17. 1

19. 0.0003

21. $0.001 = \dfrac{1}{1000}$;

    $0.0002 = \dfrac{1}{5000}$

23. $0.00025 = \dfrac{1}{4000}$

Exercises 3.2

1. 14%

3. 42.5%

5. 100%

7. 0.03%

9. 8%   13. 5%
11. 58.$\overline{3}$%   15. 83.3%

Exercises 3.3

1. 1.45 or $1\frac{9}{20}$ 17. 2.784 or $2\frac{98}{125}$

3. 2.03 or $2\frac{1}{30}$ 19. 1.5 or $1\frac{1}{2}$

5. 0.507 or $\frac{38}{75}$ 21. 3.00
                                             23. 1.26
7. 0.65 or $\frac{13}{20}$ 25. 77.5632
                                             27. 0.0002
9. 1.81 or $1\frac{57}{70}$ 29. 0.0049

11. 0.609 or $\frac{3}{5}$ 31. $0.1\overline{6}$ or $\frac{1}{6}$

13. 17.875 or $17\frac{7}{8}$ 33. 16.7%

15. 0.001 50 or $\frac{3}{2000}$

Review

1. a. $\frac{1}{20}$ 4. a. 15%
                                                b. 150%
    b. $\frac{2}{3}$ c. $0.\overline{6}$%
                                                d. 68%
    c. $\frac{63}{20}$ e. 0.02%

    d. $\frac{1}{12\ 500}$ f. $166\frac{2}{3}$%

2. a. 0.05   5. a. 0.83
    b. 0.67   b. 0.475
    c. 3.15   c. 0.017
    d. 0.000 08   d. $0.0\overline{5}$ or $\frac{1}{18}$

3. 0.0025% = 0.000 025   e. 28.$\overline{3}$

    $=\frac{25}{1\ 000\ 000}$ f. 0.01 or $\frac{1}{100}$

    or $\frac{1}{40\ 000}$ g. 1.65
                                                h. 0.025
    0.1% = 0.001
                                             6. $\frac{1}{16}$ of or 0.0625 of; 6.25%
    $=\frac{1}{1000}$

7.   2.5 times; 250%

8.   a.   1.5% = 0.015

     b. $\dfrac{7}{8} > 87\%$

c. $5\dfrac{1}{2}\% < \dfrac{11}{2}$

d.   3.7% < 3.7

Chapter 4

Exercises 4.1

1. $\dfrac{1}{4}:\dfrac{7}{8}$ or $\dfrac{2}{7}$

3.   500 milligrams : 1 capsule

    $\dfrac{500 \text{ milligrams sodium citrate}}{1 \text{ capsule}}$

5.   50 000 units : 1 tablet

    $\dfrac{50\,000 \text{ units drug}}{1 \text{ tablet}}$

7.   100 units : 1 millilitre

    $\dfrac{100 \text{ units Regular Iletin}}{1 \text{ millilitre}}$

9.   100 milligrams : 1 millilitre

    $\dfrac{100 \text{ milligrams drug}}{1 \text{ millilitre}}$

11.   10 grams : 150 millilitres

    $\dfrac{10 \text{ grams}}{150 \text{ millilitres}}$

13.   1 cup : 1 quart

    $\dfrac{1 \text{ cup sugar}}{1 \text{ quart solution}}$

15.   15 milligrams : 1 square metre

    $\dfrac{15 \text{ milligrams Dimethane}}{1 \text{ square metre}}$

17.   50 micrograms : 1 kilogram

    $\dfrac{50 \text{ micrograms Choloxin}}{1 \text{ kilogram}}$

19.   2 litres : 18 hours

    $\dfrac{2 \text{ litres solution}}{18 \text{ hours}}$

21.   1200 millilitres : 24 hours

    $\dfrac{1200 \text{ millilitres dextran (6\%)}}{24 \text{ hours}}$

23.   5 parts sodium chloride : 100 parts whole solution

    $\dfrac{5 \text{ parts sodium chloride}}{100 \text{ parts whole solution}}$

Exercises 4.2

1. $x = 9$
3. $x = 1$
5. $x = 16.2$ or $16\dfrac{1}{5}$
7. $x = 45$
9. $x = 35$
11. $x = 33$
13. $x = 0.41\overline{6}$ or $\dfrac{5}{12}$
15. $x = 4$

17. $x = 0.5$
19. $x = 7.5$
21. $x = \dfrac{1}{28}$
23. $x = 10$
25. $x = 29.1\overline{6}$
27. $x = 0.6$
29. $x = 4$
31. $x = 36.\overline{36}$

Exercises 4.3

1. $2\frac{1}{2}$ tablets sodium bicarbonate (600 mg/tab)

3. $1\frac{1}{2}$ tablets dicumarol (100 mg/tab)

5. $\frac{2}{3}$ tablet
    (750 mg/tab)

7. $\frac{1}{2}$ tablet Antrenyl (5 mg/tab)

9. 5.7 millilitres amobarbital
    (44 mg/5 mL)

11. 0.$\overline{54}$ or 0.55 millilitre
    0.55 millilitre histamine
    phosphate (275 mcg/mL)

13. 3 millilitres ferrous sulfate
    (125 mg/mL)

15. 2.75 millilitres testosterone
    (100 mg/mL)

17. a. 252 micrograms
        scopolamine

    b. 0.84 millilitre
        scopolamine
        (300 mcg/mL)

19. a. 510 micrograms
        digoxin

    b. 10.2 millilitres digoxin
        (50 mcg/mL) in divided
        doses

21. a. 0.33 gram theophylline

    b. 61.1 millilitres
        theophylline
        (0.027 g/5 mL)/24 h

23. a. 3 suppositories/24 h

    b. 1 suppository/dose

25. 1 suppository/dose

Review

1. $4\frac{1}{2}:9;\ \dfrac{4\frac{1}{2}}{9}$

2. 1 litre : 7 grams; $\dfrac{1\ \text{litre}}{7\ \text{grams}}$

3. 0.25 : 100; $\dfrac{0.25}{100}$

4. $x = 7\frac{1}{2}$

5. $x = 5$

6. give 10 millilitre
    fluphenazine (2.5
    milligram/5 millilitre)

7. a. 37.5 milligram

    b. give 0.75 millilitre
        dimenhydrinate (50
        milligram/millilitre)

8. give 1 tab (500 milligram)
    erythromycin/dose

Chapter 5

Exercises 5.2

1. $3\frac{1}{2}$

3. $3\frac{1}{2}$

5. $f\!\!\!z\ \frac{1}{3}$

7. $f\!\!\!z\ \frac{3}{4}$

9. $z\frac{1}{2}$

11. $1\frac{1}{4}$ pt

13. $f\!\!\!z\ \frac{1}{4}$

15. gr 30

17. $f\!\!\!z$ 24

19. ♏ 90

21. gal 2

23. z 1

25. qt $2\frac{1}{2}$

27. fʒ $\frac{1}{3}$

29. pt 4

31. gr 60

33. gal $1\frac{1}{2}$

35. ♏ 80

37. gr 25

39. ʒ 100

41. ʒ 4

43. ʒ 16

Exercises 5.3

1. $5\frac{1}{2}$ tablets

3. $\frac{3}{4}$ tablet

5. $1\frac{1}{2}$ tablets

7. $2\frac{2}{3}$ tablets

9. $2\frac{2}{5}$ tablets

11. fʒ $2\frac{2}{3}$

13. ♏ 10

15. ♏ $11\frac{1}{4}$

4. ʒ $3\frac{1}{3}$

5. fʒ $\frac{1}{2}$

6. ʒ 12

7. fʒ 8

8. ♏ 120

Review

9. gr 15

10. gr 240

11. 8 pints

12. give 2 tablets (gr 5)

13. give ♏ 60 or ʒ 1

14. give ♏ 10.7

Chapter 6

Exercises 6.1

1. 1000 mL

3. 1000 mg

5. 0.025 g

7. 1520 mL

9. 800 000 cg

11. 200 cm

13. 250 mm

15. 0.25 m

17. 1.5 g

19. 6 mg

21. 20 000 µg

23. 1.625 L

25. 0.0054 kL

27. 1500 mL

29. 2.5 g

31. 200 000 cg

33. 0.1 L

35. 200 mm

37. 500 mL

39. 0.28 m

41. 354 mm

43. 3000 g

45. 6.5 cm

~~47. 600 mcg~~

49. 45 000 mcg

51. 936.8 mm

53. 0.001 km

~~55. 0.250 mg~~

57. 0.05 g

59. 50 000 μg

Exercises 6.2

1. 2 dm^3

3. 2000 mL

5. 2 kg = 2000 g

7. 1.05 g = 1050 mg

9. 0.001 L

11. 3 cm^3 = 3 mL

13. 0.003 kg = 3 g = 3000 mg

Exercises 6.3

1. 2 tab (500 mg/tab), QID

3. 3 tab (500 mg/tab)

5. $\frac{1}{2}$ tab (2g/tab)

7. $\frac{1}{2}$ tab (0.3 mg/tab)

9. 2.5 mL (0.2 g/5 mL

11. 15.38 mL (650 mg/5 mL)

13. $1\frac{1}{2}$ tab Simethicone
(50 mg/tab) QID

15. 1 tab reserpine
(125 mcg/tab), BID

17. $1\frac{1}{2}$ tab methyclothiazide
(2.5 mg/tab)

19. 5 ml hydrocodone
(5 mg/5 mL), PO, TID

Review

1. 2000 mg

2. 0.05 mg

3. 1500 mL

4. 30 000 mcg

5. 10.5 mm

6. 0.015 m

7. 0.022 kg

8. 0.005 dL

9. 0.012 mm

10. 500 μL

11. 5 g

12. 5 mg

13. 5 cm^3

14. 5 kg

15. give 1 tab (400 mg), TID

16. give 0.70 mL (0.3 mg/mL)
atropine sulfate, IM, q6h

17. give 7.6 mL (6.25 mg/5 mL)
promethazine, PO, q4h

Chapter 7

Exercises 7.1

1. gr $4\frac{1}{6}$

3. 0.36 g

5. gr $12\frac{1}{2}$

7. 120 mg or 130 mg

9. gr $\frac{1}{600}$ or gr $\frac{3}{2000}$

11. 200 mcg

13. 0.005 g

15. ℥ 20

17. 1 tsp

19. 9 tsp

21. ℳ 240

23. $gr \dfrac{3}{1300}$, or $gr \dfrac{1}{400}$

25. $f \text{℥} 13\dfrac{1}{3}$

1. $36.3\overline{6}$ kg
3. 17.82 lbs
5. 30.48 cm
7. 86.36 cm
9. 6.61 in
11. 3175 μm
13. 25 kg, give 406 mg Acetylsalicylic acid, QID
15. 502 μg digoxin
17. 85.45 kg, 17 mL magnesium sulfate (50%)

1. 2 times

3. 0.5 or $\dfrac{1}{2}$

5. 10 times
7. Give one 0.2 g tablet
9. a. 1 suppository (8 mg)
   b. 1 suppository (0.12 g)

1. $gr \dfrac{1}{3}$

2. 120 mg
3. 3 tsp
4. 0.75 mL
5. 3 tbsp
6. 30 000 mcg
7. 41 kg
8. 45.72 cm
9. 38 100 microns
10. 15 mL; 3 tsp

27. $3\dfrac{1}{3}$ tbsp

29. 1 mg

Exercises 7.2

19. 0.0003 in, or $\dfrac{1}{3175}$ in

21. 0.1339 in

23. 7.1 m

25. $f \text{℥} 16\dfrac{2}{3}$

27. 2.86 lbs

29. 0.0055 in, or $\dfrac{7}{1270}$ in

Exercises 7.3

11. 1.9 mL caffeine and sodium benzoate (250 mg/mL)

13. 820 mL isotonic sodium chloride

15. 9 days at most

17. $gr \dfrac{1}{120}$ is larger

Review

11. administer 1 suppository (125 mg) theophylline, BID

12. give 1 tab (0.3 mg) atropine sulfate, PO, q6h

13. $\dfrac{1}{3}$ of

14. 2 times

15. 818 g

16. give 2 tsp (6.25 mg/5 mL) promethazine, PO, at bedtime

Chapter 8

Exercises 8.1

1. $\dfrac{\text{Novocain}}{\text{solution}}$ $\dfrac{\text{gr } 1.5}{\text{m } 100}$, $\dfrac{1.5 \text{ g}}{100 \text{ mL}}$, $\dfrac{0.000\ 25 \text{ g}}{1 \text{ mL}}$, $\dfrac{0.25 \text{ mg}}{1 \text{ mL}}$,

   $\dfrac{0.015 \text{ g}}{1 \text{ mL}}$, $\dfrac{15 \text{ mg}}{1 \text{ mL}}$, $\dfrac{15\ 000 \text{ mcg}}{1 \text{ mL}}$ $\dfrac{250 \text{ mcg}}{1 \text{ mL}}$

3. $\dfrac{\text{mannitol}}{\text{solution}}$ $\dfrac{\text{gr } 20}{\text{m } 100}$, $\dfrac{20 \text{ g}}{100 \text{ mL}}$,

   $\dfrac{0.2 \text{ g}}{1 \text{ mL}}$, $\dfrac{200 \text{ mg}}{1 \text{ mL}}$,

   $\dfrac{200\ 000 \text{ mcg}}{1 \text{ mL}}$

5. $\dfrac{\text{gr } 1}{\text{m } 4000}$, $\dfrac{1 \text{ g}}{4000 \text{ mL}}$,

7. $\dfrac{\text{m } 0.125}{\text{m } 100}$, $\dfrac{0.125 \text{ mL}}{100 \text{ mL}}$

9. $\dfrac{\text{glycerin}}{\text{solution}}$ $\dfrac{\text{f} \text{ʒ } 1}{\text{f} \text{ʒ } 20}$, $\dfrac{1 \text{ L}}{20 \text{ L}}$

11. $\dfrac{\text{m } 1}{\text{m } 800}$, $\dfrac{1 \text{ mL}}{800 \text{ mL}}$

Exercises 8.2

1. 4 mL methocarbamol (1 : 10 W/V)
3. 0.5 mL epinephrine (1 : 10 000 W/V)
5. 12.5 mL Acetaminophen (120 mg/5 mL)
7. 25 mL Novocain (1.5% W/V)
9. 201.6 mL glycerine (50% V/V)
11. 25 mg mouthwash
13. 0.35 mL epinephrine (1 : 1000 W/V)
15. 6 drops potassium iodide 100%
17. m 10

Exercises 8.3

1. 20 g drug needed to make 500 mL, 4% W/V soln
3. 2.5 g drug needed to make 250 mL, 1 : 100 W/V soln
5. 4 mL of 5% soln can be made from 0.2 g drug
7. m 240 soln (1 : 20 W/V) can be made from gr 12 drug
9. 2% W/V soln
11. a. 40% W/V
    b. 2 : 5 W/V
    c. 2 g/5 mL
13. 500 mL; dissolve 0.5 g drug in sufficient diluent to make 500 mL solution. Label 0.1% W/V.
15. 4 g sodium perborate are needed to make 200 mL solution. Label sodium perborate 1 : 50 W/V.
17. Dissolve 50 g Epsom Salt in water sufficient to make 125 mL solution. Label Epsom Salt 40% W/V.

Exercises 8.4

1. 75 mL acetic acid (6%) is needed. The 75 ml acetic acid (6%) is slowly poured into 75 mL distilled water to make 150 mL acetic acid (3%) soln.
3. 30 mL 0.1% soln needed; to 30 mL (0.1%) soln add liquid sufficient to make 150 mL, which will be 0.02%.

5. 8000 mL of 0.0025% soln can be made from 20 mL of 1% soln.

7. 333 mL of 3% soln can be prepared.

1. $\dfrac{gr\ 1.5}{m\ 100}, \dfrac{1.5\ g}{100\ mL}, \dfrac{0.015\ g}{1\ mL},$

   $\dfrac{15\ mg}{1\ mL}, \dfrac{15\ 000\ mcg}{1\ mL}$

2. $\dfrac{gr\ 1}{m\ 5000}, \dfrac{0.02g}{100\ mL}, \dfrac{0.0002\ g}{1\ mL},$

   $\dfrac{0.2\ mg}{1\ mL}, \dfrac{200\ mcg}{1\ mL}$

3. $\dfrac{60\ mL}{100\ mL}, \dfrac{m\ 60}{m\ 100}$

4. $\dfrac{1\ mL}{5\ mL}, \dfrac{f\, \backslash\!\!3\ 1}{f\, \backslash\!\!3\ 5}$

5. give 200 mL (0.5% W/V soln)

9. 1 mL 5% cupric sulfate is needed; dilute with diluent to make 200 mL soln.

6. give 4 mL (0.75% W/V soln)

7. give 3.2 mL (4%) lidocaine

8. give 0.26 mL (1 : 1000) epinephrine

9. 25 g

10. 0.1 mL

11. use 5 mL of 1% soln and add diluent sufficient to make 10 mL

12. 480 mL of 1 : 4000 W/V soln

Review

Chapter 9

Exercises 9.1

1. Syringe setting: 60 unit level. Draw U-100 Regular Iletin into a 100-unit insulin syringe to the 60-unit level.

3. Give 0.60 mL of U-100 Regular Iletin in a 1-mL syringe.

5. Syringe setting: 25 unit level. Draw U-40 isophane insulin suspension into a 40-unit insulin syringe to the 25-unit level.

7. Give 0.50 mL of U-500 Regular Iletin in a 1-mL syringe.

Exercises 9.2

1. Give 0.50 mL sodium penicillin G (1 000 000 units/10 mL).

3. Give 1.3 mL heparin sodium (15 000) units/mL).

5. Give 0.68 mL bacitracin (5000 units/mL), BID.

7. Give 1.1 mL penicillin G potassium (250 000 units/mL), QID.

9. Give 1.7 mL bleomycin 10 units/mL.

Exercises 9.3

1. 1 cap vitamin E (100 units/cap)

3. 2 tab penicillin G potassium (250 000 units/tab), QID

Review 1. a. Draw U-100 Regular Iletin into a 100-unit insulin syringe to the 70-unit level

          b. Administer 0.70 mL of U-100 Regular Iletin

     2. Draw U-40 insulin into a 40-unit syringe to the 25-unit level.

     3. Give 0.54 mL of U-500 insulin via a 1 mL (tuberculin) syringe.

     4. Give 0.22 mL (5000 units/mL) bacitracin, IM, BID.

     5. Give $1\frac{1}{2}$ tab (200 000 units/tab) penicillin G benzathine, PO, QID

Chapter 10

1. a. 46 gtt/min if using 10 gtt/mL delivery system.

    b. 69 gtt/min if using 15 gtt/mL delivery system.

    c. 92 gtt/min if using 20 gtt/mL delivery system.

    d. 275 gtt/min if using microdropper delivery system.

3. a. 8 gtt/min if using 10 gtt/mL delivery system.

    b. 12 gtt/min if using 15 gtt/mL delivery system.

    c. 17 gtt/min if using 20 gtt/mL delivery system.

    d. 50 gtt/min if using 60 gtt/mL delivery system.

5. 27 gtt/min if using 15 gtt/mL delivery system to finish on time.

7. a. 2720 mL to be delivered in 14 h.

    b. 194 gtt/min if using microdropper delivery system.

9. a. Total time to administer 2000 mL is 405 min (about 6.75 h).

    b. 296 gtt/min via microdropper delivery system to deliver 2000 mL dextrose 10% in 6.75 h.

Chapter 11

1. Using Fried's rule, 3 mg Nembutol.

3. Using Young's rule, 100 mg tetracycline.

5. Using Clark's weight rule, 15 mg gentamicin.

7. Using Clark's body surface rule, 120 000 units penicillin G.

9. Using Formula 5:2.28 g sulfisoxazole.

Chapter 12

1.

	Anterior	Posterior
Trunk	3.5	4.3
Buttock	2.5	2.5
Perineum	1.0	
Rt thigh	3.25	3.25
Rt leg	2.5	2.5
Rt foot	1.75	1.75
Lt thigh	3.25	3.25
Lt leg	2.5	2.5
Lt foot	1.75	1.75

Total = 43.8%

3.

	Anterior	Posterior
Head	1.17	
Neck	5.0	2.5
Rt arm	1.0	1.0
Lt arm	0.75	0.4
Trunk	9.7	2.16
Perineum	1.0	
Rt thigh	1.58	
Rt leg	0.7	
Rt foot	0.58	
Lt thigh	2.0	
Lt leg	0.8	
Lt foot	0.58	

Total = 30.9%

Index

*Page numbers in italics indicate terms
that are defined in the glossary.